PRIMEVAL CREATURES
OF THE ANIMAL WORLD

CONTENTS

Published in 1993 by Thomasson-Grant
World copyright © 1992 Edizioni White Star,
Via Candido Sassone 22/24, 13100 Vercelli, Italy

Any inquiries should be directed to:
Thomasson-Grant, Inc., One Morton Drive, Suite 500
Charlottesville, Virginia 22903-6806
(804)977-1780

Photography credited on page 144.
Printed in Italy

99 98 97 96 95 94 93 5 4 3 2 1

Library of Congress
Cataloging-in-Publication Data available
ISBN 1-56566-039-O

PAGES 2-3

Iguana, Galápagos Islands

PAGES 4-5

Crocodile, Nile River

PAGES 6-7

Great white shark, Great Barrier Reef, Australia

PAGES 8, 9

Mountain gorilla, Zaire

PAGES 10-11

African elephants, Kenya

PAGES 12-13

Humpback whale, Gulf of Alaska

Preface
Fulco Pratesi

Text
Marco Ferrari

Edited by
Valeria Manferto De Fabianis

Designed by
Patrizia Balocco

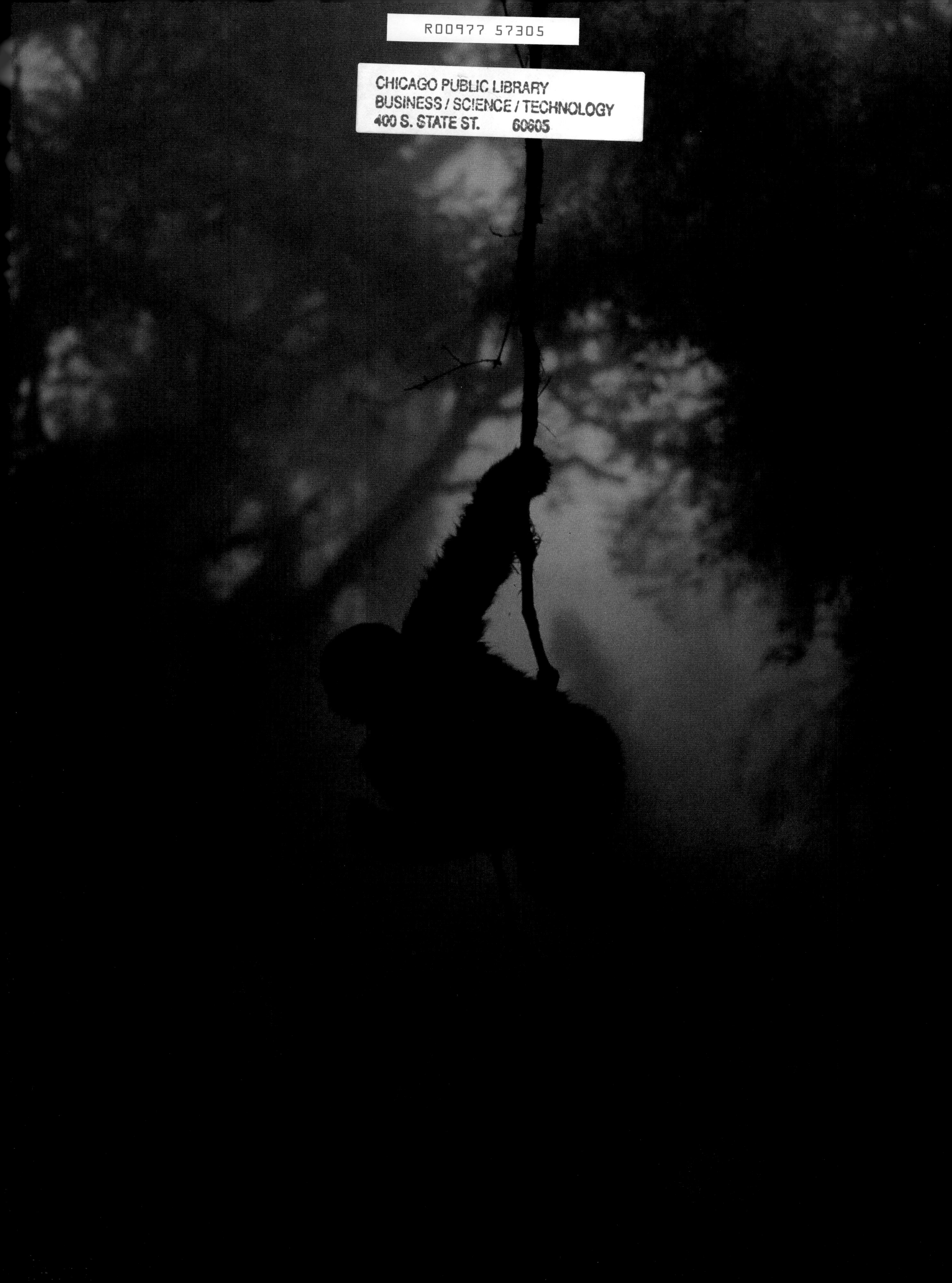

The Fascination of Prehistory

As a child I was first introduced to the world of nature by Figuier, a naturalist of the nineteenth century. From the yellowed pages of his marvelous book came forth dark visions of primordial landscapes. I recall romantic illustrations of swamps of the Carboniferous period filled with giant ferns, vast prairies crossed by ferocious dinosaurs, a horizon of volcanoes in the smoke of which flew pterodactyls with bat wings and crocodile snouts, and Jurassic oceans bristling with fangs, horribly elongated necks, and monstrous fins.

The Earth before the Flood, I think it was entitled. Figuier gave a bewitching idea of how our planet must have looked before that immense catastrophe, represented as an enormous flood submerging the Earth, with overhead a leaden sky streaked by thunderbolts and flashes of lightning. From then on, the charm of prehistory has been not only always at the back of my mind, but sometimes lying tangibly in my hands. I once saw a series of sharks' teeth, still bright and clear, held fast in conglomerate on Italy's Mount Majella, now the realm of wolves, eagles, and roe deer.

At an altitude of 4,000 feet in a wood-encircled valley in Abruzzo National Park, I was magically transported into an environment like the Australian barrier reef, teeming with life, when I realized that the stones before my eyes were the remains of ancient, brain-shaped corals. Small rods protruding from the stones were none other than sea urchin spines identical to the ones I had seen in the fissures of living coral in the Maldives. Thus, all at once, beech trees and moss, the call of the jay and the flight of the raven were supplanted in my imagination by the image of an ancient tropical lagoon where butterflyfish and moray eels, surgeonfish and mantas swam among the waves.

Another time, a relative who is an amateur paleontologist discovered the huge femur of a prehistoric mammoth while wandering in a gravel quarry on the outskirts of Rome. I recall the care with which he cleaned and packed this fragment of sandy stone that had belonged to one of the largest land mammals ever to have existed. And once, while on a birdwatching ramble in a gorge near the Via Aurelia outside Rome, I discovered, in a newly dug ditch, a curved and cylindrical "stone" jutting out of the clay. I immediately thought of an elephant tusk because of the area I was in, and I was greatly surprised when university experts informed me that the enormous find didn't belong to a pachyderm but to a wild ox (one of the terrible aurochs Virgil mentions in the Georgics) which grazed in the region during prehistoric times.

Finally, to come to eras closer to our own, I remember a game I used to play as a boy in which we rummaged in a quarry of diatomaceous earth that had been dug in the Tiber Valley, not far from Rome. Flakes of this soft white rock were easily opened with a well-aimed hammer blow. Often, though not always, they revealed the treasure they had preserved since the dawn of time: impressions of the leaves of hornbeam or linden (similar to those we find today), but above all, small fish, perfectly clear on both sides of the stone, probably fallen to the bed of the ancient swamp and there placed in the archives between diatom skeletons.

But no dinosaurs. In my part of the world few traces of these gigantic, terrifying reptiles remain. Other countries hold deposits where we can admire the footprints of these animals and find traces of the Cretaceous period, the period in which the dinosaurs reached their maximum point of expansion before disappearing in an immense catastrophe. This was not the great flood of Biblical memory, of course, but another punishment inflicted on the planet by an enormous meteor that darkened the atmosphere for centuries with dust raised by its impact. The dust produced a long winter capable of annihilating the colossal reptiles. At least this is what is claimed by those who base their theories on the layers of iridium in Cretaceous-era rocks all around the world.

I saw no dinosaur fossils as a boy, but today, as illustrated in this book, we may discern primeval life in the forms still present among us. To see how, we must first recount the history of life on Earth.

In the first two billion years of our planet there was no trace of life as we understand it. The first rocks erupted from the depths of the Earth, boiling avalanches of lava and magma spread out over the surface, mountains and volcanoes were born and annihilated in terrible explosions, and in their interior, there was nothing suggesting any

The high level of humidity in tropical forests prevents perfect fossil formation—very few remains have been found in these rich environments. However, it is not improbable that in the past the forests were, as they are now, a perfect habitat for reptiles. Iguanas are the modern representatives of a group that dominated the Earth for millions of years, the dinosaurs, and they still maintain some of the characteristics of their massive ancestors. The basilisk (above) of Central and South America, peculiar in appearance, also has the remarkable ability to scurry across the surface of the water on its hind legs.

(RIGHT) Dorrigo National Park, Australia

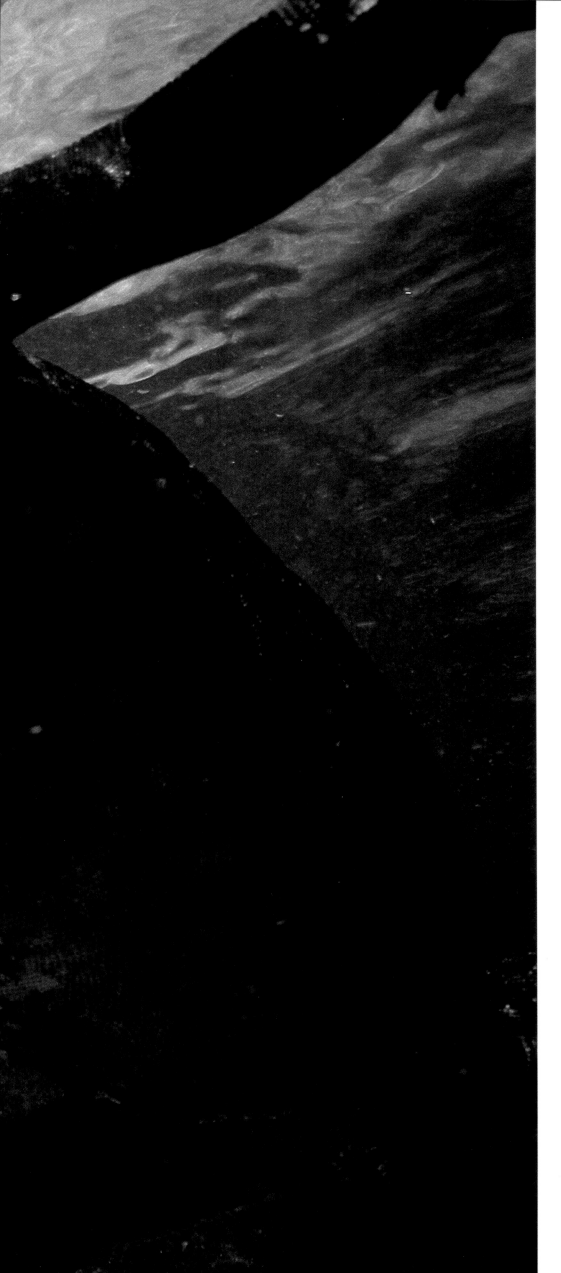

The marine iguana of the Galápagos Islands is the only vegetarian reptile which feeds in the sea, where it harvests seaweed. Its carnivorous ancestors reached these remote islands either by swimming or by catching rides on floating tree trunks.

form of life whatever. Yet, even then, from the red-hot and teeming cauldron of the primary oceans, electrical discharges, mixtures of gases, and why not, the finger of God, brought about the miracle: a transparent, evanescent, ephemeral molecule endowed with the admirable capacity of reproducing itself. From this molecule, in millions of years of convulsions, thunderbolts, and chemical reactions, the first real, albeit rudimentary cell was formed, and millions of years later, the first groups of living cells, too fragile to leave any fossil trace.

Then, about a billion and a half years ago, the first organized creatures arrived—monocellular algae swimming in the thick soup of the Cambrian seas. Later, geometrical trilobites, worms, and jellyfish evolved. But it was only about 600 million years ago that the first creatures of a certain size and with characteristics not too different from those we now know began to populate the Earth and leave their fossil footprints.

Which were the ancestors of the plants and the animals now on Earth? The algae, as we have seen, colonized marine water right from the primordial period; it is thanks to the oxygen released by them that life was possible. The next plants adapted to a subaerial life seem to have been Cooksonia, no more than two inches high and with no leaves, roots, or fruit. Their dispersion was entrusted to spores contained in a spherical sack.

Just as for plants, the roots of the animal family tree are sunk into the mire of the ocean beds. As far as one can deduce from a study of the fossil record, it was about 500 million years ago that the first "fish" arose. This was a rudimentary creature with a body covered with bony plates like a medieval knight. Since it didn't have jaws, it crawled along the seabed sucking up particles of food. This ostracoderm, as it is known, became extinct during the Devonian period and was replaced by placoderms, marine animals whose fins and scales better equipped them for survival. The successors of the placoderms in the tumultuous depths were other creatures which can be considered the first step forward of the animal kingdom towards a terrestrial existence. Crossopterygii fish, very similar to the coelacanths discovered this century in the waters of the Comoro Islands, had lobed fins and a semblance of lungs. Over a period

17

lasting millions of years between catastrophes, they were able to take the first hesitant steps on land, where oxygen produced by plants had already created a livable atmosphere.

This was 365 million years ago. In the rocks of Greenland dating back to this period, geologists have discovered the remains of an animal that can be considered the transit point between fish and amphibians and the ancestor of modern salamanders (in the same way as the ostracoderm was the ancestor of modern fish). With its well-developed and articulated legs, its efficient, tooth-filled jaws, and its more or less functional lungs, this moist, crawling creature finally managed to leave the cauldron of the prehistoric oceans and begin its conquest of the planet's shores.

There was no shortage of food. Already thirty million years previously, the first insects had left aquatic life and followed the invading march of the plants onto dry land. Ichthyostega, an apparent attempt by natural selection to obtain a model of a terrestrial animal, succeeded. Fossil remains of these animals, in different forms, have been found in Europe, North America, and some parts of Asia.

Naturally, the march of the land animals could not stop with the ichthyostega. The next step produced by the marvelous process of evolution was the development of eggs enclosed in shells much tougher than the defenseless, gelatinous eggs of fish and amphibians. Being able to protect the embryos outside the protective liquid element gave a new independence to these creatures. The oldest fossil egg to have been discovered dates back 280 million years and was found in Texas.

The passage from the amphibians to the reptiles, which were to become dominant for millions of years, was laborious and interminably long: the intermediate models, still half amphibian and half land animal, were innumerable. One of the creatures that can be considered as one of the most likely ancestors of the reptiles was the hilonomus, which appeared on Earth about 300 million years ago and disappeared once more into the mists of geological time about 100 million years later. It was a saurian similar to a lizard and about two feet long.

The immense and spectacular dinosaur family invaded all parts of the globe. Most notable were such famous characters as the enormous apato-

saurus, the terrible tyrannosaurus, the awkward triceratops, and the slender iguanodon. There were of course many others, large and small, carnivorous and herbivorous, terrestrial or entirely aquatic. Some species followed a new path: pterodactyls, having modified their front limbs to hold a thin membrane of skin, took to the skies. Others, which did not descend from the former, developed plumage and elongated their limbs to form real wings, becoming the first precursors of birds. Known as archaeopteryx, they appeared 140 million years ago.

What about the mammals? From the boundless selection of reptiles of all shapes and sizes another branch sprouted. A group of small, not very noticeable creatures set out on its evolutionary march along unbeaten tracks. With successive attempts, the ancestors of mammals managed to overcome the climatic and terrestrial conditions that led to the disappearance of the dinosaurs. Their limbs were longer, thinner, and better placed with respect to the body, and they had more efficient thermoregulation facilitated by a covering of hair insulating them from heat and cold. They also had systems of reproduction that did not require the production of external eggs and guaranteed protection by means of a placenta, making these primitive creatures well-suited for survival.

One of the many progenitors of mammals as we know them today was probably the creature cynognathus, which lived 250 million years ago. It was still a reptile, but had several mammalian characteristics. Its skeleton and cranium were similar to those of a dog, its legs were placed below and not at the side of the body (as they still are in modern-day saurians), it had vibrissae (whiskers like a cat) on its face, and perhaps also had fur. This model was one of the many to be discarded.

And man? Fortunately, man was the last to appear on the world's stage. From australopithecine man, whose first fossil remains date back 2.8 million years, to us, the human species has undergone important modifications. The path has been long and fascinating since the day monkeys, pushed by necessity or by changed climatic conditions, left the forest in search of prey in the prehistoric savanna. They acquired in the course of millennia an erect stance, binocular vision, prehensile upper limbs,

The depths of the sea teem with life forms whose fundamental structure dates back millions of years. The nautilus (above) is the only living cephalopod with an external shell. It is virtually identical to its ancestors which lived 400 million years ago, having managed to survive all the great extinctions of the past. The fossil record shows that jellyfish (right) have remained virtually unchanged from the time of their first appearance. The acquisition of efficient weapons, stinging cells called nematocysts, has enabled these primitively structured animals to continue to populate the seas.

continuous growth of the cerebral mass, and finally, the use of complex tools, which has brought civilization from the first stone utensils to modern electronic computers.

This admirable saga, which starts with the first monocellular blue algae wandering in the oceans and doesn't stop at the reader of this volume, has, naturally, left many traces on our planet. As in any story worth its salt, shreds and relics of bygone epochs, archaeological remains and souvenirs are strewn along the path of life.

Fossils—skeletons and carapaces, shells and footprints—are found everywhere from the Dolomites in Italy to the bottom of the Grand Canyon in Colorado. Geologists and paleontologists use the ages of fossils to date the various rocks and determine evolutive processes in one territory or another. However, over and above fossils, other precious relics have survived to our day in both the plant and animal kingdoms.

Let's start with the plants. The beautiful tree ferns of the rain forests or the cycads of tropical jungles are not very different from the ferns of the Jurassic period. Drops of resin that became amber in the course of millions of years probably came from conifers not so very different from those found in the Alps. And there are plants such as horsetails in today's swamps that closely resemble the plants appearing in reconstructions of Carboniferous landscapes.

The same is true for fish. The coelacanth of the Indian Ocean is a true copy, with virtually no variations, of fish thought to have become extinct long ago. The elegant and geometrical shells of argonaut and nautilus cephalopods recall the immense ammonites that populated the sea in the Triassic period. Observing the large horseshoe crabs on North American beaches, how can we fail to notice their similarity to the trilobites of the Permian period?

For the reptiles, naturally, given the large number of species from which they descend, there are many examples of "primeval creatures." Perhaps most admirable is the sphenodon of New Zealand, almost identical to the homeosaur, a Jurassic reptile belonging to the Rhynchocephalia order thought to have become extinct at the end of the Cretaceous period. Sharks, crocodiles, and turtles, not

only because of their archaic appearance, are a tangible reminder of their ancestors of millions of years ago. And among the birds, how can we forget the strange hoatzin of the South American forests, a living fossil closely resembling in its anatomy, appearance, and life cycle prehistoric birds which became extinct millions of years ago?

Coming to periods closer to our own and to ancestors more closely related to us, we have the famous duck-billed platypus, which lays eggs like a bird, but, as a mammal, suckles its young, Others include the tree shrew, the spiny echidna, and of course, the primates—our great great grandfathers and great uncles. Gorillas, orangutans, and chimpanzees, although they belong to different branches of our complicated genealogical tree, are closely related to us and represent our closest link with nature, a link we seem to be doing our utmost to forget and neglect, cutting away the very roots of our existence.

The book you are sitting down to enjoy gives a splendid idea of all that evolution in the course of hundreds of millions of years has left on our planet. Our aim is to make you realize the extreme importance of each of those tiny pieces of the puzzle we call life.

Fulco Pratesi

Despite its curious and primitive appearance, the hoatzin of South America is a highly evolved bird with complex adaptations. Its particular habits and a diet of fruit and leaves have transformed this hen-sized bird into a replica of the ancestor of all birds, the archaeopteryx. The hoatzin has a short and labored flight, the wings of its nestlings have very small claws, and its feathers are ruffled and loose. It was feathers, perhaps "invented" as heat regulators but soon transformed into organs for flight, that enabled birds to diversify and exploit almost all the Earth's habitats.

PAGES 22-23
Volcanic eruption, Hawaiian Islands

PAGES 24-25
Rain forest canopy, Papua New Guinea

INTRODUCTION
Sparks of Life

Until a short time ago, words like *primitive* and *primordial* suggested incompleteness and crudeness. However, now we have begun to rediscover the charm of a primitive past in which everything was less complicated. "Primordial" takes us back to remote times before the birth of man, amid hovering mists and ancient forests, when dreadful dinosaurs and fierce predatory mammals ruled the Earth. What remains of these times are only traces of forests increasingly under siege, vast seas, and wastelands of ice and sand. The intention of this book is to discover the primordial, the untouched, the places where nature still reigns supreme, where ecological equilibria are still respected and biological richness has not been diminished—a sort of blessed state, in which animals and plants live side by side.

Only now are we beginning to discover the impact and importance of symbiosis in every ecosystem. The criterion this book uses to illustrate this vision of the world is essentially ecological. We have divided the Earth's surface into five large vital domains: the sea, wetlands, savannas and deserts, mountains and polar regions, and forests. And it is in these environments that we have gone to seek out the characteristic species that create the atmosphere of a "lost world."

At first sight the sea may be considered an almost unitary environment, but very slight changes in temperature and irradiation can cause drastic modification in ecosystems. For the forests, the biomes covered to a great extent by tall trees, the problem exists in even more radical terms. There are at least three main forest subdivisions whose conditions vary greatly with a change in latitude. There are widespread northern forests, dominated by conifers, showing very few variations from one part of the world to another, and deciduous forests, previously widespread at middle latitudes and most suffering from the impact of human transformation. Most complex of all, we have the tropical forests, the richest and most vital ecosystems on Earth.

At the other extreme we have placed the deserts, including both savannas and steppes, which, in the strictest sense, are not deserts. These are the habitats where we find the large hoofed animals and the best known predators: lions, hyenas, leopards, and cheetahs. In the section dedicated to the mountains we have placed the extreme environments of the poles and the tundra, for obvious reasons of similarity. Finally, the wetlands, which cover a smaller area but are equally important, also include apparently marginal but fundamental zones such as resting places for migratory birds and breeding grounds for innumerable species.

With these points in mind, we welcome you to join us in a short but indispensable excursion to discover the wealth of life on this planet we continue to exploit without knowing thoroughly.

Marco Ferrari

Facing difficult climatic conditions and a great scarcity of food, only a few species can survive in extremely cold environments. Polar bears (above) are among the most successful inhabitants of cold mountain and polar regions. When Europe was covered by the snows of the ice ages, some species of bear competed with man for dominance. Now they have been driven into marginal environments, and it is only in the Arctic that the bears can hunt with relative tranquility; their main prey are the seals which emerge from holes in the pack ice.

(RIGHT) *The Japanese macaque's woolly coat and its habit of sticking close to thermal hot springs help it survive in a cold climate.*

PAGES 28-29
Indian rhinoceros, Royal Chitwan National Park, Nepal

CHAPTER ONE
The Seas

ater has always been inextricably linked with life. The very properties of the molecule (in particular its great capacity as a solvent and its efficiency as a filter against ultraviolet radiation and changes in temperature) were necessary for the origin and evolutionary development of living beings. Even today water is present in variable quantities in all animal and vegetable species and is indispensable for life as we know it.

The evolutionary history of life in this liquid began about 3.5 billion, perhaps 3.8 billion, years ago. The planet had just cooled down, and already complex chemical reactions between organic molecules formed the first building blocks of life. These interactions were guided by well-known physical and chemical forces; their action occurred however on the basis of principles still not well known which led to an increase in the complexity of systems.

As chemical systems became more complex, molecules arose which were able to replicate themselves. Water enabled these fragile compounds, little more than long chains of amino acids, to "live" and interact in an increasingly intricate way between themselves and with the environment without being degraded or transformed. The passage from molecule to a simple cell with a nucleus (and eventually to groups of cells) seems to have occurred soon thereafter.

The enormous period of time from the start of the history of life on Earth has enabled species living in water, and in the seas in particular, to diversify and evolve ever more complex systems. For example, at the beginning of the Cambrian period (about 570 million years ago) hundreds of different species originated in a period of time which, geologically speaking, was very short. These belonged to all the major groups we know today as well as to others that have now disappeared.

During the Cambrian, the first laws of ecology were established. The producers, the plants, were food for the primary consumers, the herbivores, and in their turn these were preyed on by the carnivores, the secondary consumers. Efficiency in capturing energy (because basically photosynthesis and predation are none other than methods of capturing the sun's energy) increased as time went on, together with the intricacy of reciprocal relationships.

As well as cradling the origin of life, the sea was the starting point for the majority of living species. Even those which originated and diversified on land, like reptiles and mammals, have representatives in the sea. Chelonia (the turtles) among the former and cetaceans (whales and dolphins) among the latter, have returned to the sea, even though it cannot be said that the turtles are completely marine. In fact, they must return to land to lay their fragile eggs.

The true rulers of the sea are the fish. Their ancestors "invented" jaws by transforming a branchial arch (one of the bones supporting the gills) into an extraordinary instrument of nutritive success. The difference in efficiency and offensive power between a lamprey, which does not have jaws, and any other fish is enormous. There are perhaps 30,000 species of fish (broadly speaking), more than all the other vertebrates put together. Their adaptations and their ability to survive in extreme conditions are extraordinary; they vary greatly in size, from the gigantic whale shark to the tiny males of deep ocean fish that spend their entire life as parasites on the much larger females of the species. Many species can withstand extremes of temperature, salinity, and pressure.

Among the fish, we find the most extraordinary examples of "living fossils," species having remained virtually unchanged for tens or even hundreds of millions of years. The coelacanth is the best-known example, but also many cartilaginous fish (sharks and rays) are not very different from their ancestors that lived about 400 million years ago. A shark may be said to attack its prey not only from the depths of the ocean, but from the depths of time.

Sharks (together with less well-known but equally fascinating creatures like the ratfish) are classical examples of successful groups. Some sharks' physiology and anatomy have evolved to such sophisticated levels that very few other vertebrates can surpass them. Yet their physical form has remained essentially similar to the form of primitive vertebrates.

The fascination of the sea does not derive solely from the individual species found in it, even if these are the result of an evolution that has lasted for billions of years. It is their union in an

The sea is the biome with the largest number of apparently primitive animals whose close ancestors date back millions of years. One species can certainly be called a living fossil—the coelacanth (right). Discovered by Europeans (Comoro Island fishermen had known about the coelacanth for some time) in 1938 on the ship Nerine *by the curator of the East London Museum in South Africa, this strange fish has all the characteristics of its ancestors. Four hundred million years ago, these lobe-finned fish differentiated themselves and established a line that led to the amphibians. The horseshoe crab (above) also has a timeless structure; its form is very similar to that of the trilobites of 400 million years ago.*

apparently infinite number of ecosystems that attracts scientists, tourists, and "fish-watchers" every year. The oceans cover about seventy-one percent of the Earth's surface and, apart from the particular case of the open ocean, they are also characterized by enormous differences in conditions which increase the number of ecological niches. Between one sea and another there can be differences in salinity, temperature, pressure, the concentration of nutrients, and consequently the number of species and the composition of ecosystems. The greater the number of physical settings in which species can hunt, hide, and lay their eggs, the greater are the "work opportunities" (ecological niches).

The relationships among living species (predation, symbiosis, parasitism) increase with the variety of habitats. The richness of an ecosystem is defined by the number of such interactions. One of the fundamental characteristics of marine ecosystems is the presence, side by side, of structurally primitive and extremely evolved organisms in competition. For example, the large sharks were present, in forms very similar to those of today, in the seas of the Jurassic and Cretaceous periods along with marine reptiles, plesiosaurs, tilosaurs, and icthyosaurs. In the seas of today, the white shark (*Carcharodon carcharias*) probably competes for its prey (mainly squid and octopus) with the sperm whale, a highly evolved mammal. In the same manner, along rocky coasts one can find simple sea anemones and sea urchins side by side with very specialized crustaceans. A single dive into the sea, especially near a coral reef or a rocky coastline, can serve as a lesson in the increase in complexity brought about by evolution. The range extends from jellyfish, equipped with an elementary body design, to annelids, relatives of the common earthworm, to sea urchins and starfish, on up to crustaceans, fish, and mammals. There are few ecosystems so diverse as marine ones.

There are two main variables determining the differences between the various marine environments: solar irradiation (which influences the temperature and photosynthesis of the phytoplankton in particular) and the presence of nutrients. When both are high, as in shallow tropical seas, conditions exist for one of the most complex environments of the entire planet—the coral reefs. In these laboratories of evolution, hundreds of animal and plant species live together in an intricate series of relationships dating back millions of years. Here every species seems bizarre and surprising. The stability of these environments, their great wealth of nutrients, and their high temperatures make them "lost worlds," where everything takes place according to primeval laws still not completely understood.

Extensive studies on the animal communities of the coral reef have clarified some of the fundamental principles of ecological science. Coexistence without competition between species of fish has been explained on the basis of the different types of niches they occupy. Although they are apparently similar and live side by side, some fish feed on coral, others on plankton, and still others on encrusting seaweeds. Some species survive on a diet of parasites they find on the scales of other fish. Here, the most incredible shapes and the brightest hues have a precise function of determent or mimicry. It is thus not surprising that about 6,000 to 8,000 species of fish (about thirty to forty percent of all bony fish) are associated in one way or another with the coral reefs.

Other ecological communities are simpler but equally fascinating. In the cold waters of the Arctic and the Antarctic, powerful sea currents cause thousands of tons of nutrients (mainly phosphorous and nitrogen) to rise to the surface. Here, billions of living beings (phytoplankton, zooplankton, and larger crustaceans) create a simple but very important food chain sustaining the marine beings that are probably the most fascinating and evolved of the Earth—the cetaceans. Whales and dolphins, ecological if not taxonomic descendants of the large marine reptiles that dominated the shallow seas of the Mesozoic period, are at the peak of the food chains in almost all the seas. Large and efficient predators, they have not yet completely revealed to man the enormous complexity of their hunting methods, social interactions, and other behavior patterns. Schools of whales crossing entire oceans are composed of individuals that know each other personally and communicate with each other in languages unknown to us, perhaps "speaking" from one part of the ocean to the other.

Mantas, like sharks, are carti-laginous fish. Their shape derives essentially from the habits of their ancestors, who adapted to life on the sea bed by enormous expansion of their

pectoral fins. This allows for a strong swimming motion in a less conspicuous, flatter form.

PAGES 36-37
Ligurian coast, Italy

It is hard to believe that these immense animals evolved from species with relatively banal characteristics, not very different from a dog—the mesonychidi. In what is a short time in evolutionary terms, several tens of millions of years, these small predators assumed completely different forms. Their bodies lengthened and their hind legs disappeared, while their front legs were modified into long fins ideal for swimming. But the most spectacular transformations made breathing possible for animals living almost constantly underwater, and ensured their extraordinary underwater delivery of young, their capacity to hear underwater, and their re-adaptation to a diet of plankton. In this way, a land animal gradually became completely aquatic and began to dominate the Earth's seas.

The most bizarre living forms, true primeval creatures, are those living in places where sunlight never shines. In the deep ocean, strange beings with lengthened appendages, atrophied or hyper-sized eyes, with colonies of luminous bacteria on the surface of their bodies, attract other fish, often feeding on prey larger than themselves. A few alienlike species are the superpredators of this system. The coelacanth, perhaps the most famous "living fossil," was discovered by Western man near the Comoro Islands in 1938. One of the largest sharks, the so-called megamouth (*Megachasma pelagios*), was first captured and described in 1976, and only two examples are known.

All the ecology of the deep seas cannot directly depend on sunlight. Bodies of dead animals and waste coming from the surface fall constantly to the bottom and are rapidly broken down by bacteria. In such a way simple food chains are established at great depths. One of the most remarkable marine ecosystems was discovered in the 1980s. Around underwater vents of hydrogen sulfide at very great depths, a rich bacterial flora develops, giving sustenance to enormous annelids and other species, some of considerable size. These ecosystems are completely independent of the need for solar energy and exist only on the sulfurous compounds issuing from the bowels of the Earth. Despite decades of research, the oceans still harbor many secrets, and may still hide strange and terrible creatures yet unknown.

Nutrient-rich coral reefs like the Great Barrier Reef of Australia (left) are the marine environments with the greatest variety of life. This wealth derives in part from their incredible antiquity—the first corals date back to the origin of multicellular organisms, about 600 million years ago. Traces of calcium carbonate deposited by coral have been found in most parts of the world. For millions of years, open-sea predators like the barracuda (left) have used coral reefs as a hunting ground. Fish school in confusing, swirling patterns (right) in response to a threat by such predators. Fossil beds containing hundreds of fish bear witness to the antiquity of this behavior.

PAGES 40-41
Whale shark, Red Sea

All sharks are carnivores, but not all are dangerous to larger fish or mammals. Some species, like the basking shark (below) and the largest shark, the whale shark (upper and middle right) have become filterers, feeding on plankton and very small fish. The remains of the first cartilaginous fish date back 360 million years, and numerous groups of sharks, some of them living in fresh water, have succeeded them in the course of time. Modern sharks such as nurse sharks (lower right) first appeared about 60 million years ago. The most famous of the fossil sharks, Carcharodon megalodon, rivaled the large plankton-eating sharks in size, reaching a length of forty feet.

Mantas, members of the Raji-
dae order, are perhaps the most
dramatic looking rays. Carti-
laginous fish evolved at the
same time as bony fish, but the
appearance of their more
evolved cousins did not cause
them to disappear from the
oceans.

With the passage of time,
sharks have diversified into
hundreds of species that occupy
all saltwater environments. The
hammerhead's peculiar flattened
head reveals its kinship to the
family of rays.

The last to arrive in the family of ocean dwellers, whales immediately established themselves at the peak of the food chain in all the seas of the world. It was only some 50-70 million years ago that the whales' ancestors, a group of small hoofed carnivores, returned to the sea, undergoing the rapid transformation that suited them for life in the water. Starting from predators like protocetus, about eight feet long, and zeuglodon, about forty feet long, they arrived at the larger representatives like the humpback whale (right) and blue whale, which can grow to 100 feet in length. There are two major groups of whales, toothed and baleen, the latter sharing with mantas (left) the nutritive strategy of straining the sea for plankton.

PAGES 48–49
Scorpionfish, Red Sea

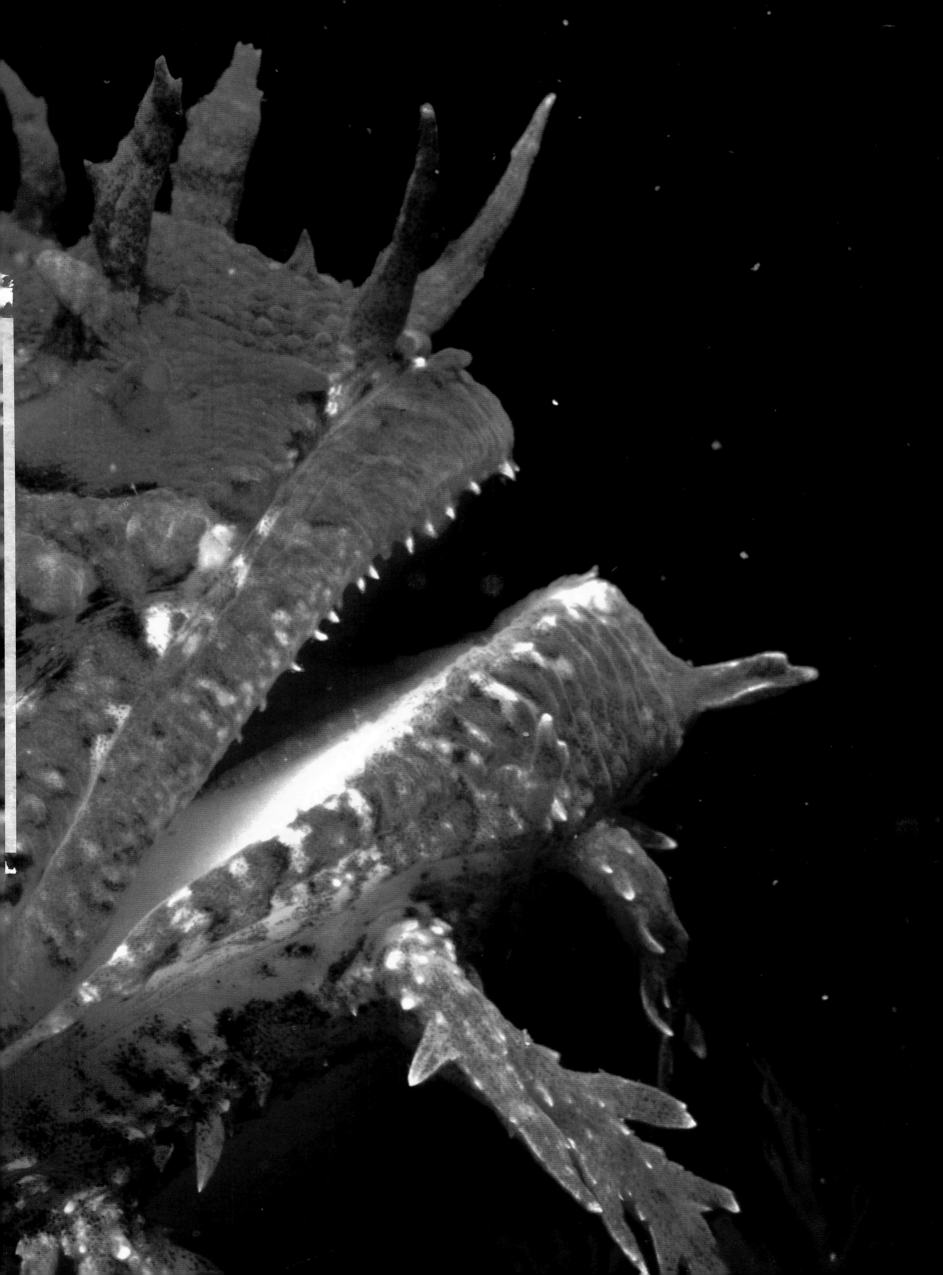

Solitary atolls and volcanic
islands often lack predators,
which has enabled many animals to remain virtually
unchanged from their primitive
ancestors. It seems that marine
iguanas, endemic to the Galápagos Islands, are much more
ancient than was once thought;
they have been separated from
their terrestrial relatives for at
least twenty million years.
These miniature dinosaurs
(above, right, and on pages
52–53) use their crested spines
and scale-covered heads in terri-
torial fights.

CHAPTER TWO

Wetlands

Endless canebrakes, thick fogs, trees as bare as ghosts, evil-smelling miasmas, mosquitos, malaria: this is what we think of when we hear the word *swamp*. Only six percent of the Earth's surface is covered by swamps and other wetlands, and the animal and plant species that live there are not as spectacular as those living in other environments, but they are no less important or interesting than other terrestrial biomes.

Wetlands are only apparently marginal ecosystems. Their lack of points of reference and their hybrid nature (neither water nor land) do not make them easy to understand. The human race is not known as a friend of swamps—during the 1980s, half of the world's wetlands were drained. Wetlands are subjected to swift and continuous evolution. Detritus fills the basins, new species replace previous ones, and the landscape itself has a different appearance every year, adjusting to climatic conditions very quickly. Because of these hard conditions, wetland species sometimes need extreme and very interesting adaptations.

We apply the word *wetland* to many different habitats, sharing a single feature—the preponderant presence of shallow and rather stagnant, soil-permeating water. So lakes, estuaries, rivers, and swamps may be included under this definition. The water may be fresh, salt, or brackish, and either rich or poor in oxygen.

Climatic and edaphic (of the soil) conditions generate wetlands with varying degrees of nutritive elements: vast African inland lakes, temporary and very changeable; river deltas, perfect shelter for extraordinary fauna and flora; temperate and northern estuaries, where thousands of animals migrate for breeding; and North American potholes, resting and refuge sites for ducks and geese. Perhaps most peculiar of all are mangrove swamps, where highly specialized plants send their roots into the water without suffering from the presence of sea salt.

These ecosystems, set between water and land, have elements of both. The fascination they hold for us and their importance for the species of the Earth are due to this mixture. Each latitude and climatic condition generates swamps, ponds, and bogs with different characteristics. Therefore, the species living in each habitat are different, although it is often possible to notice common physiological features between plants and animals living in continents far from each other.

In Arctic areas environmental conditions (strong winds, cold rains and snows, and frozen soil) dwarf the growth of trees. Scraps of forest grow in the small islands of dry land created by the wandering of water, and the great difference between the seasons causes cycles of abundance and shortage. Animals spending summer on the tundra take advantage of these cycles, from flights of millions of birds to hoofed animals in herds once numbering in the tens of thousands. The large water birds, swans and cranes, most strike the human imagination. Their mating dances and white plumage are among the most magical sights of a visit to the tundra. And in the small lakes that border on conifer forests are found the loons, whose strange singing may be said to symbolize the northern tundra.

At lower latitudes, water basins on the coast are often joined to the river estuaries that flow sluggishly to the sea. They form large, shallow, muddy expanses. Ecological conditions are extremely changeable during the year, but in summer, when the water is warmed by the sun for many hours a day, algae—the basis of almost every food chain—undergo a huge explosion in population. Tides bring an uninterrupted flow of nutritive elements, changing estuaries in summer into ideal places for birds to find food and regain energy lost during migration. The result is that thousands of birds, in particular shorebirds such as killdeers and sandpipers, synchronize their migration patterns in order to find themselves in the right place at the right moment. They use these wetlands as resting and breeding sites during spring and summer. Birds then migrate to the tundra, which is also covered by a film of water. Moreover, they use estuaries to restore fat before their long journeys towards the south. During these periods flocks of millions of birds cover the sky and swirl in unison over the muddy land, creating breathtaking spectacles.

Temperate and subtropical wetlands provide wintering and resting sites for European and American ducks during the cold season. Even in winter the number of birds is amazingly large. The clearest example of the importance of wetlands for fauna is the presence of thousands of

In prehistoric times, swamps covered a much larger expanse than they do now, representing one of the largest environments on the surface of the Earth. In the meandering stretches of slow-moving rivers some species of fish, growing increasingly similar to amphibians, conquered the terra firma. In the course of time and thanks to the absence of predators on the land, they perfected lungs and legs, which were more suitable for a life out of the water. However, amphibians like the crested newt (above) still needed water to facilitate reproduction.

(RIGHT) *King Sound, Western Australia*

waterfowl floating in the Mediterranean or Central American swamps, that is to say in rather unfriendly environments. Farther north, in the temperate zones, human expansion has left few places that can be defined as swamps. Among these are the estuaries of wide rivers such as the Camargue in France, the Coto Doñana in Spain, the Danube Delta in Romania, the shores of the North Sea, and the wide lakes and bays bordering the east coast of the United States. The flyways of the North American interior make up a major system of wetlands; along these migratory routes, thousands of ducks and geese stop to reproduce.

Flooded forests, where trees grow directly out of the water, are to be found mainly in tropical areas. Relatively constant temperature and conditions allow some tree species to adjust to the chronic lack of oxygen in stagnant waters. The spectral, moss-laden cypresses, for instance, in North American swamps such as the Everglades, capture oxygen in the atmosphere by means of roots rising out of the water.

In overflow areas situated in the deltas and along the banks of wide rivers (the Mississippi, Amazon, Mekong, Congo, and Irrawaddy), strips of flooded forest are created, and the clear distinction between aquatic and terrestrial animals and plants becomes indistinct. Some species, including the herons and kingfishers of Indonesian forests, lead their life cycle on dry land, but feed in water. Amphibians must lay their eggs in water and go back on land when they are fully grown.

In Africa, a series of tectonic lakes (having their origin in the shiftings of the Earth's crust) give shelter to at least a dozen species of herons, storks, and sparrows, six species of ducks, and two of flamingos. These birds are to be found in Lakes Naivasha, Natron, and Nakuro, grouped in flocks comprising tens of thousands of birds.

Mangrove swamps offer a good example of very rich and threatened wetlands. Along the coasts of tropical seas fairly constant climatic conditions are favorable to the growth of about twenty species of trees belonging to different classes (Rhizopora, Sonneratia, Ceriops, Avicennia, etc.) generally known under the common name of mangroves. They survive despite great variations in tide and salinity, mainly because their tangled roots trap the

sediments carried by the tides, allowing mangrove swamps to grow toward the sea.

Such an unpredictable, various environment calls for very special adaptation. Consequently, throughout the history of the Earth, natural selection has favored those plants and animals that could live both in water and on the land. In shallow subtropical waters, in the marshes, and in the temporary lakes of the Devonian period (between 410 and 360 million years ago) continuous changes in water levels forced some fish to develop organs similar to our lungs. It was during the Upper Devonian (perhaps even before) that in shallow marshes bordered by the giant tree ferns, the first real terrestrial vertebrates, amphibians, came out of the water.

Many fossils belong to the Ichthyostega class, an intermediate stage between fish and amphibians. Even the development from amphibians into reptiles occurred in a marshy environment. The innovative evolution of an egg independent from water took place in shallow marshes during the Lower Permian, when the first strong predatory reptiles (pelicosaurus and other similar beasts) replaced amphibians as dominant terrestrial carnivores.

A leap in time of millions of years allows us to observe life in the swamps of the Mesozoic. During this era the supercontinent Pangaea split, and shallow seas separated the continents; in this environment dinosaurs, which can be regarded as the most important inhabitants of prehistoric swamps, thrived. Of course, only a few among the nearly 900 species of the two dinosaur classes (Saurischia and Ornithischia) lived in damp areas, swamps, estuaries, or shallow inland seas. During the Mesozoic, ecological productivity was very high, and therefore many herbivores were attracted by the plentiful nourishment of the wetlands. As a consequence, carnivores looked for their prey there. Given the warm climate and the shallow seas (actually salt or brackish marshes), it is easy to understand why the evolution and development of some "terrible lizards" during the Jurassic and the Cretaceous was strictly dependent on wetlands.

The theory maintaining the necessity for some dinosaurs to live in the water in order to support their enormous weight has been discarded. Nevertheless many fossils prove that there was

Frogs, including this Amazonian horned frog eating a grasshopper (right), though primitive in appearance, are very different from the amphibians that populated the land in the Upper Devonian period. They are much smaller and much more physiologically complex.

PAGES 60-61
Submerged forest, Northern Europe

PAGES 62-63
Badwater Pool, Death Valley, California

an adaptation from marsh-living animals into terrestrial ones. Camasaurus, for instance, had nostrils on the upper side of its head, allowing it to breathe while keeping under water; androsaurus and lambeosaurus probably fed on tender marsh grass. Moreover, tracks on ancient lake shores show places where predatory pteropods might have hidden waiting for their prey. Contemporary with dinosaurs, but not of the same family, pterosaurs and pteranodons, with their long beaks and tails, flew about high in the sky looking for small fish or tiny crustaceans to take into their filtering beaks. Surprisingly similar to their modern counterparts, enormous crocodiles laid ambushes for watering animals.

According to ecologists, much further research must be carried out to give a satisfactory explanation of the ecological transformation of and succession in the world's wetlands. It is in impenetrable and flooded forests and in the dead oxbow lakes of the Congo basin that cryptologists look for the descendants of the biggest inhabitants of swamps, those "lost dinosaurs" still so appealing to us. And in some unknown oxbow lake there might be a living fossil, hidden by fogs and among mosquitos, which could give an explanation for other mysteries of evolution.

Swamps shelter many species with ties to primordial times. Manatees (upper left and at right) are the only aquatic mammals that feed on grass. Their origin can be traced back to animals which were no larger than a dog and very similar to the modern-day hyrax. The hippopotamus (lower left), which started out as a land-dwelling herbivore, has become increasingly dependent on water. Roaming the Earth for millions of years, crocodiles (below) adopted a style of predation which enabled them to survive the massive extinctions at the end of the Cretaceous period.

While not reaching the fifty-foot-length of their ancestors in the Triassic period, crocodiles are still one of the largest existing reptiles. Armed with a set of highly efficient teeth, they have substituted the prey of the Mesozoic period with zebras and other animals that wander too close to the water. Because of their relatively unchanged anatomy, crocodiles have often been used for studies aimed at discovering the nature of the dinosaurs. Some modern theories positing that dinosaurs kept a constant body temperature are based on studies of the crocodile.

The swamps at the foot of the Himalayas provide refuge for some very uncommon animals and are the undisputed realm of the Indian rhinoceros. These beasts are the last descendants of a once more numerous group of rhinoceroses. Under their armor they have certain similarities to horses. Once widespread throughout the planet, rhinoceroses were in ancient times a successful group with representatives in tropical and temperate zones. They had very large horns, and their enormous bodies protected them from the cold. The evolution of a thick skin and withdrawn style of life defended them from

their enemies, but hindered their development. These animals now live in wet regions where frequent mud baths help them get rid of skin parasites. A curious fact is that male Indian rhinoceroses fight using their teeth rather than their horns.

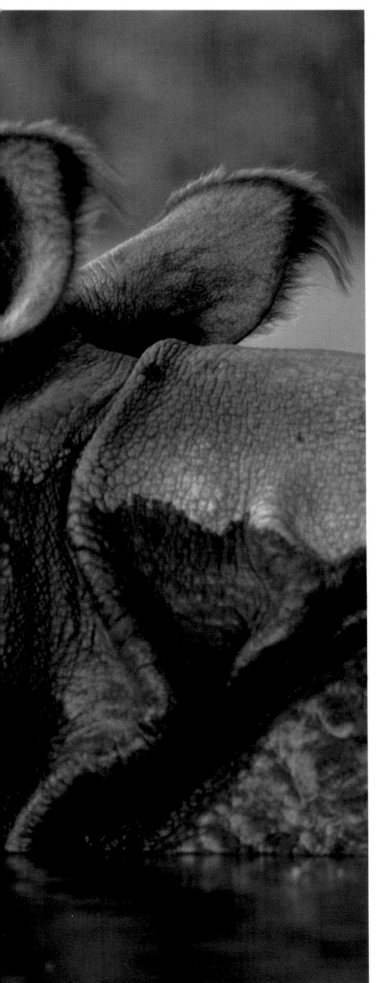

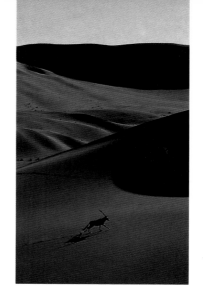

CHAPTER THREE

Grasslands, Savannas, and Deserts

If there is any truth in sociobiologist Edward Wilson's hypothesis, then it would seem that in the course of its expansion throughout the world, the human species has preserved the memory of its ancestral environment. This original "home" would appear to have been rather similar to modern grasslands. Indeed, it was in the African savannas—enormous stretches of grass dotted with a few umbrella trees of the acacia family—that the adventure of the first hominids began. Unlike the vast areas of land planted with grain for human consumption, natural grasslands are biologically diversified ecosystems with a wide range of characteristic plants and animals.

The factor which conditions life in the grasslands and prevents the development of forests is the amount of rainfall. One could say that grasslands, at least the tropical ones, represent the halfway house between two important ecosystems—forests and deserts. As research proceeds, it becomes increasingly clear that grassland ecology, apparently elementary and similar all over the world, is actually exceedingly complex. The interaction and competition between grasses and trees, the different extension of root systems, the enormous herds of herbivores living on the grasslands, the intricate recycling of nutrients, and last but not least, the impact of human activities, have created a mosaic of environments still largely to be discovered.

An ecosytem often considered together with the grasslands, but which is not perfectly identical, is the savanna. Characteristic in the savannas of tropical regions is the presence of tall trees and bushes interspersed among the dominating grasses. Savannas cover about twenty percent of the Earth's land surface; they are found in mountainous regions, on the plains, and on a great variety of soils.

The African savannas are always in an unstable equilibrium between becoming deserts in dry years and transforming themselves into dense forests during heavy rainfall. The dominant vegetative form is grass: there are dozens of species, mainly Gramineae, which, with extraordinary adaptations, manage to survive critical periods and germinate when there is sufficient humidity. However, every grassland has a few species of dominating grasses, ranging in height from five feet to the twelve feet reached by elephant grass.

At times, seeds and the rhizomes of Gramineae manage to wait for tens of years in a state of quiescence. They can also survive unharmed the frequent fires (often caused by humans) which represent one of the most important factors in the modeling of herbaceous ecosystems. In the dry season the roots of grasses are more efficient than those of trees in the underground struggle to reach water. The flowering season is short but intense; after a torrential rainstorm an expanse of dry grass is transformed for several days into a flowering meadow.

The trees dotting the savannas have to enter into competition with grasses, and thus they have very deep roots. Their foliage is often wide and flat to defend their roots from the burning sun. Both trees and bushes produce an enormous quantity of seeds; the karoo acacia of South Africa, for example, can produce up to 20,000, ninety percent of which are fertile.

The largest system of temperate grasslands lies in the so-called paleoarctic region of Europe and central and northern Asia. This vast expanse is more correctly referred to as a steppe, which derives from Russian and means "grassland devoid of trees." It is not easy to imagine the wealth of animal life once flourishing in these enormous environments. The monotony of the landscape conceals a very important fact—the extremely high efficiency of photosynthesis. The grassland transforms the greater part of sunlight into energy in the form of herbaceous matter, which in its turn, nourishes, or nourished, an almost unimaginably large number of grazing herbivores.

Unfortunately human action has overtaken all of the Euro-Asian ecosystems, and of the enormous herds of hoofed animals that roamed the steppes (saiga, bison, wild horses), many exist only in prehistoric cave paintings. Along with the large herbivores, predators like wolves and eagles have also disappeared. There remain only small herbivores and their predators: woodchucks, hamsters, and voles and the polecats and snakes that creep into their lairs.

The great North American prairie was not very different from the prehistoric steppe. Before the massacres of the 19th century, there were fifty to sixty million American bison in the central

Over time, the territory covered by deserts has varied considerably. When the continents were all united in one huge land mass, the interior, distant from the sea, received almost no precipitation. This led to the creation of the first and perhaps largest desert on Earth. The split-up of the original supercontinent, Pangaea, reduced the surface area of the desert and distributed small deserts around the globe. Then, according to the increase or reduction of temperature or the occurrence of glaciations in our era, the deserts retreated or increased in size. In every epoch, however, only very specialized species have managed to survive in this difficult habitat.

(ABOVE) *Oryx, Namibian Desert*
(RIGHT) *Queensland, Australia*

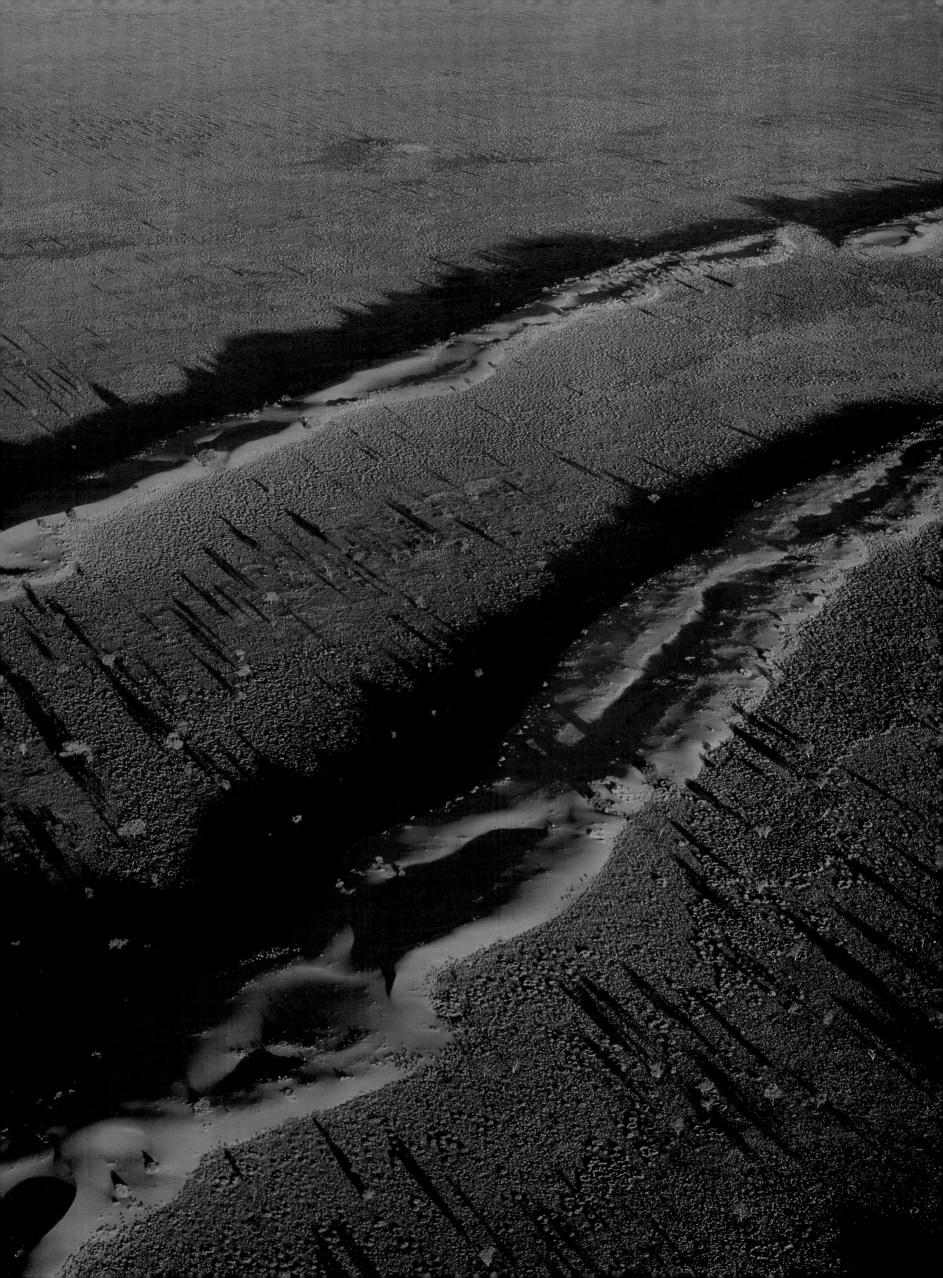

prairies (which, ironically, were probably created by the first men who entered the American continent about 12,000 years ago). Together with the bison there were a large number of antelopes as well as numerous prairie dog towns. Prairie dogs' tunnel systems once extended over areas of several square miles. One must not forget the millions of grasshoppers and other herbivorous insects which provided nutrition for many species of insectivorous birds. All these animal species, in a continuous interaction with the environment, contributed to maintaining the structure of the prairies. Bison and antelope nibbled the grasses continually, keeping them low, and ate the few shoots of trees surviving in the harsh climate. All the charm of these environments is now limited to a few American and Asian national parks. The only predator which has managed to withstand the action of man is the coyote, an opportunistic member of the Canidae family not terribly fussy in choosing its prey.

In the southern hemisphere there are less well-known systems of temperate grasslands. In southern Brazil, Uruguay, and Argentina, the pampas were probably created by the fires started by the first inhabitants coming from the North. The conditions of extreme aridity and high temperature, combined with almost constant winds, are such that very few animals manage to survive throughout the entire year. Many birds migrate, and some mammals spend the majority of their time under ground. Rare predators like the beautiful maned wolf and the pampas fox lie in ambush waiting to capture the elusive rodents.

Although its area is smaller than the Euro-Asian steppes, the African savanna is much more famous. Perhaps because they recall our past as "monkeys descended from the trees," the grassy expanses of eastern Africa still impress us. They compose the only environment on Earth where it is still possible to study large concentrations of herbivores and their predators. According to a modern theory, the large African hoofed animals, differently from the American and European ones, have managed to survive because their cohabitation with the most efficient predator, man, has been much longer, thus making reciprocal adaptation easier.

The different seasons impose extremely precise life cycles on hoofed animals and an incessant search for water and food. Wildebeest, zebra, antelope, elephant, and their predators—lions, leopards, cheetahs, and hyenas—compose the classic image of life in the savanna. Because of the wide-ranging ecological studies carried out in the course of more than a century by researchers from all corners of the globe, the African savanna is the environment we know best. The interrelationships between the animal and plant populations, the presence of predators, the action of fire, and the periodic dry and wet seasons in the life of the savanna form a complex fresco in which every species occupies a special niche.

It is in the savanna that mammals reach the largest dimensions and create the most complex social organizations. Herds of elephants, prides of lions, and families of baboons have been the subject of long-term studies that reveal the existence of refined social behavior. As in the temperate grasslands of long ago, the efficiency of photosynthesis on the savanna sustains life for a large number of herbivores. Size and speed are the two main strategies chosen by the herbivores for survival. Few predators would dare attack a herd of elephants or a mother rhinoceros to kill the young, and no carnivore (save the cheetah, which can reach remarkable speeds over short distances) is likely to catch a gazelle when it is running at full speed.

Such a great quantity of animals could not fail to attract many predators. Each uses different hunting strategies to exploit its prey. Lions form groups that apparently come to an agreement about assault tactics; hyenas and other wild dogs follow their prey until they drop from exhaustion; and large vultures, of which there are eight species in Africa, and other necrophagous animals pick up the leftovers from this banquet. The herbivores make an evolutionary move (increase in size or muscle efficiency, or improvement in group surveillance systems), and the predators respond with a countermove. A real arms race takes place every day in the African savanna.

In some areas, the most famous of which is the Serengeti ecosystem, grass is not constantly present, but follows the course of rainfall; the herbivores must move if they want to eat. Zebra, gnu, and gazelle gather in huge numbers and set off on great migrations, one of the last spectacles of the

The large size of the herbivores roaming the plains and savannas led to an increase in size for carnivores like lions. However, despite its size and strength, a lion must collaborate with other members of the pride to capture its prey. Grasslands have always contained large predators and large herbivores. In fossil deposits found at Rancho La Brea, California, remains of elephants and horses have been found alongside the bones of predators like the famous saber-toothed tiger. They were all trapped in wells of tar which existed in the area in prehistory. In the distant past there were also packs of predators—it is thought that carnivorous dinosaurs like deinonychus hunted in packs like lions.

Photosynthesis takes place extremely rapidly and efficiently in grasslands, making these ecosystems high in nutrition and capable of sustaining large populations of animals. Ostriches and warthogs share survival strategies on the dangerous open plains—like many animals of the savanna, they are dull colored for camouflage, and they group together for protection.

power of nature we can still observe. Around the immense herds the predators move, looking for opportunities to attack the weaker, defenseless animals, but the majority reach their goal.

The large number of habitats in the African savanna has brought about an extreme diversity of herbivore species, an evolutionary explosion which has given rise to particular habits and food preferences. Gnu and gazelle are the most striking species, but we cannot ignore the impala living on the grass left by zebra, or the marvelous greater kudu, managing to survive in the zones most densely populated by man. Extreme alimentary specialization is to be found in the giraffe, which can feed higher than any other animal, and in the tiny dik-dik that live on the shoots of thorny shrubs.

When a savanna changes as a consequence of climate or human action, a radically different environment arises. If rainfall falls below a certain level, or if man deforests or cattle overgraze the environment, the grasses begin to lose their dominion over the ground. Through a ever-narrowing spiral, grasses diminish, leading to lower soil fertility and lesser regrowth, and the ground becomes devoid of its vegetative covering. Subsequently, semiarid zones are created, rich in thorny bushes with tiny leaves. The loss of water is reduced to a minimum by the structure of the leaves, which are covered by a waxy cuticle that reduces perspiration. Plants gain some security through their root systems, often extending for many yards underground. But the impossibility of living side by side leads to excessive competition for water, and very often the larger bushes are surrounded by almost barren areas of ground; because of toxic chemical compounds that these bushes secrete into the soil, other plants, even grasses, the "children" of the bush, can't manage to germinate. The next step, especially when man introduces herds of goats into this zone, is desert, which arrives in the course of a few years.

Obviously there are also "natural" deserts, whose creation does not depend on human intervention. These occur mainly in zones in the interior of continents where the rains rarely fall or in the shadow of high mountains that block the clouds. Differently from what one might imagine, not all deserts are hot places covered with unstable and arid sand. The Gobi Desert in Central Asia, for example, is a typical cold desert, where temperatures never reach the highs found at the center of the Sahara. A fundamental characteristic of all deserts is the large gap between day and night temperatures. Even in the hottest desert, the Atacama in Chile, the nights are ice-cold. The heat of the day is not held back by vegetation, and the open terrain immediately loses the warmth accumulated during the day to the atmosphere.

This means desert adaptation in animals must ensure the maintenance of body temperature both by day and by night. During the day they must not waste too much water by sweating, because although this would lower their temperature, it would also lead to a loss of this precious liquid. At night, on the other hand, they cannot stay out in the open too long because they could easily fall prey to the many carnivores frequenting the deserts after dark. As always happens in the animal world, species must reach a compromise between obtaining food and water and avoiding predators.

In the deserts' harsh conditions, many animals have adapted similar mechanisms for survival, from physiological ones enabling them to extract the few drops of water present in seeds to anatomical ones like the large ears of the desert fox and the fennec, which aid in thermoregulation. North American rattlesnakes and North African horned vipers both travel over the sand by making s-shaped movements, a process known as sidewinding.

Plants, too, have adapted to life in the desert—some of them can survive on the small quantities of water they obtain from morning dew. The high metabolic rate of birds prevents them from colonizing true deserts. A notable exception is the Namaqua sand grouse, which has an extraordinary adaptation enabling it to slake the thirst of its young even if its water source is up to sixty miles away. The belly feathers of the male have a structure similar to a sponge that can absorb up to one-and-a-half ounces of water. Even after a long flight, the remaining water is sufficient to quench the thirst of the nestlings. Some birds of prey, and in particular those belonging to the species *Falco biarmicus*, very similar to the peregrine falcon, survive in the desert because they feed on live prey which contain quantities of water.

Bats, shown swarming out of a cave at evening, first appeared 55 to 60 million years ago. Insect-eating bats perfected complex echolocation systems early in their development.

Unlike their ancestors that pre-
ferred wetter habitats, many
reptiles now live in desert areas.
An example is the frilled lizard
of the Australian outback,
which was forced into this

inhospitable zone by the evolu-
tionary explosion of mammals
about 60 million years ago.

PAGES 82-83
Zabriskie Point, California

However, the desert is the kingdom of the reptiles. Their thick, scaly skin prevents water loss and the possibility of becoming sluggish in periods of extreme heat or cold, and this means that they can colonize arid zones, feeding on insects, seeds, or each other.

The fact that the deserts are expanding is not a piece of news to be welcomed with joy. The habitat which derives from "desertification" is even poorer than that of the desert itself. The resulting ecosystem, created by the action of man, is completely out of equilibrium. Sometimes, however, the climatic conditions can change in the opposite direction. Water returns and the grasses slowly begin to grow and colonize desert regions, creating savanna. There are examples of this around all the world's deserts—new environments that are arid and nutrient poor, but nonetheless territories which have escaped the bite of icy nights and fiercely hot days. Once the savanna has been created, normal succession can begin once more.

However, the "savanna-type" ecosystem is not constant; changes are brought about by climate, rainfall, and even the animals themselves. One of the most important ecological agents in the savanna is the elephant, which roams over almost all of sub-Saharan Africa in large herds. This animal's constant search for food and water has modified the savanna in the course of the centuries and often has been the cause of its preservation. Indeed, only the sturdiest or most indigestible trees can withstand the elephant. Elephants create an environment where the bush is sparse and grass is plentiful.

Another ecosystem-changing animal, much less visible than the elephant but according to some researchers even more important, is the termite. These small insects construct very large nests above and below ground. Their continuous movement and the displacement of enormous quantities of material, achieved in protected pathways out of the glare of the sun, contributes to the fertilization of the soil, the aeration of the top layers of earth, and the diffusion of grass and tree seeds. A landscape dotted with termite nests is only apparently tranquil—the work done by earthworms in temperate climates is accomplished in the savanna by extremely efficient termites.

In India and Australia there are environments rather similar to the African savanna. In India, the ecosystem has been created by humans, whose farming and stock-raising has destroyed preexisting forests. For this reason the species inhabiting it are few and much sought after as prey. Among the more interesting animals are Axis deer and nilgai, while the predators—lions, cheetahs, and wolves—have completely disappeared. What strikes one about the Australian savannas that surround the large central desert and border on the more humid coastal zones is the strangeness of the animals and at the same time their ecological similarity to our more classical herbivores. Here it is not antelopes or deer which nibble at the grass, but kangaroos. The predators are, or rather were, so similar to our wolf as to earn the name marsupial wolf. Here, too, as in India, human modifications have been so radical that it is virtually impossible to find anything of prehistoric size and complexity in the savanna.

The savannas and the grasslands are not only complex worlds in constant change; their history is very important for evolution. The savanna, particularly the African one, was a sort of cradle and nursery for mammals (and for the most evolved of them, man). One can claim that the forests and the swamps were the realms of the reptiles while the savannas and the grasslands were the domains of the mammals. During the Miocene epoch the large herbivores left the forests to exploit the nutritional possibilities offered by the grasslands. It was here that larger and ecologically dominant forms evolved; not only modern-day elephants, but also the primitive Cuvieronius and the giant sloths, beavers, and elephants that have become extinct. The grasslands were also home to gigantic brontotheres, with their enormous horns, as well as the largest land mammal, the indricotherium.

But these beasts, like many since them, are no more. In light of all the changes to the savanna wrought by man, we can say that the day the primitive australopithecine abandoned the protective covering of the forest to look for a better life in the grasslands was a decisive one for the future of the Earth.

The areas that are now desert were once covered with dense vegetation and supported much richer animal populations. As the deserts formed, the need to find something to eat in these arid, unproductive regions transformed reptiles such as the bearded dragon lizard of Australia (above) into efficient and speedy hunters.

The most conspicuous and important fossil deposits are to be found in deserts such as the Gobi, in Mongolia, or the western deserts of North America, where no vegetation covers the remains that have emerged.

(LEFT) *Death Valley, California*

A hard covering of scales and spines makes these reptiles look like medieval knights. However, the American horned lizard (below) and the Australian bearded lizard (right) developed their armor only for defensive purposes. Many desert species look like small versions of the reptiles which lived millions of years ago.

In the African savanna elephants represent an important force for change. The elephant family once roamed all over the world in many different shapes and sizes. Some experts claim that there have been 352 species of Proboscidea, including those which are now extinct. Some of them had tusks which pointed downwards; others had bony scales jutting from their mouths. Drastic climatic changes and competition with more efficient herbivores, combined with persecution by man, have reduced the order to only two species, represented by a few thousand surviving African (below) and Indian elephants.

Elephants' trunks and tusks are extremely efficient instruments for obtaining food and defending against predators. Few modern-day carnivores dare approach a herd of elephants to capture a calf. During the Miocene epoch in

North America, giant saber-toothed tigers preyed on elephants, but probably only when they were in evident difficulty, trapped in tar wells or close to death.

PAGES 92-93
*Wildebeest in migration,
Mara River, Kenya*

Hyenas and vultures eat carrion and scavenge behind larger predators. However, the hyena is also an efficient predator which often hunts and captures prey. Like all modern carnivores, hyenas descend from primitive Miacoidae, replacing less-evolved carnivores with less complex social lives, the Oxyaenidae and the Hyaenodontidae.

PAGES 96–97
Marabou stork, South Africa

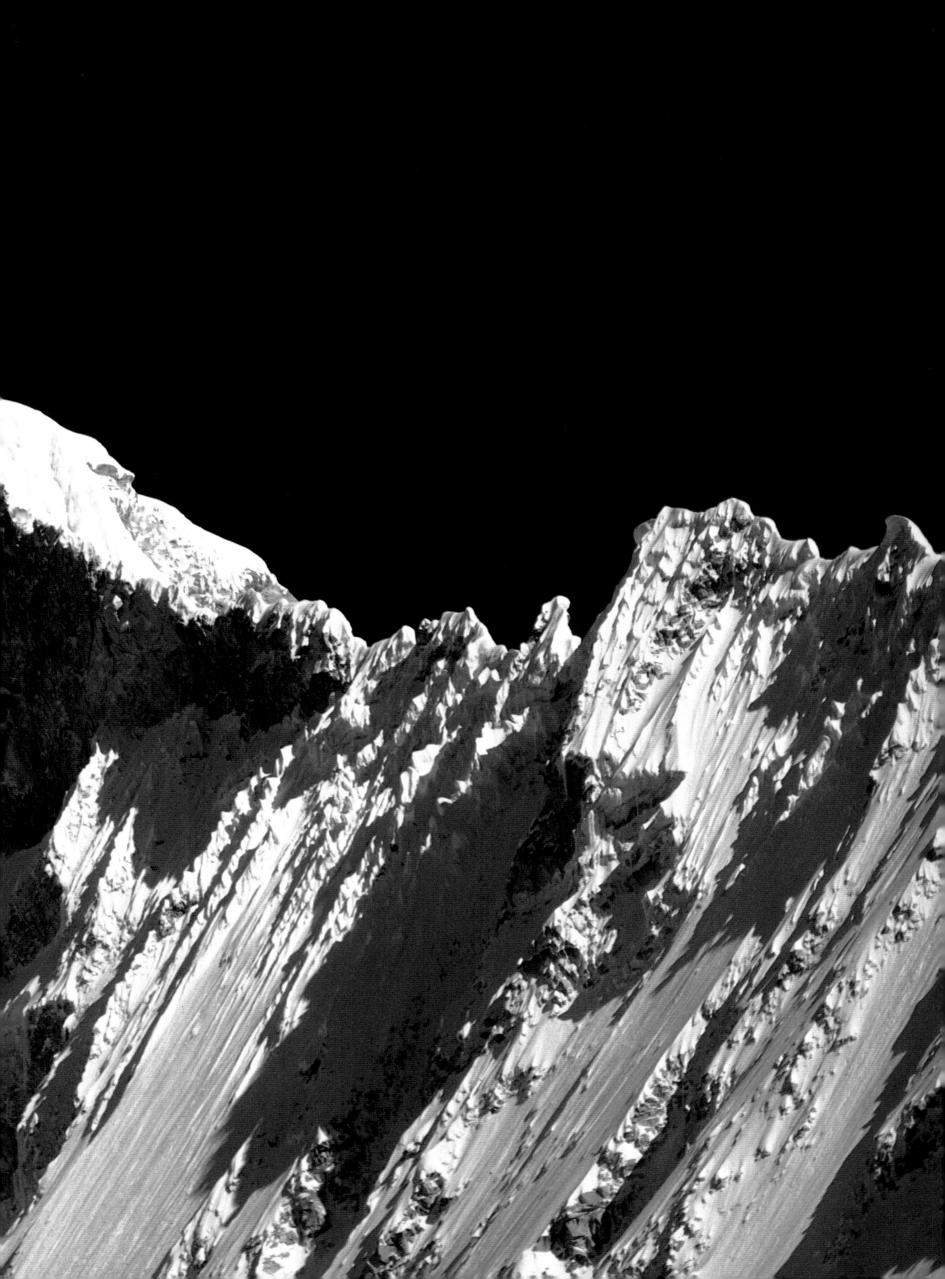

Seals and penguins show how evolution facilitates aquatic life in formerly terrestrial animals. Seals only come ashore to establish their territory and reproduce. Both descend from terrestrial ancestors (the seals from four-footed carnivores and the penguins from birds similar to the albatross and petrel), and they have had to reinvent adaptive solutions to improve movement in the water. Seals have lost the capacity to run on land, but have acquired the ability to stay underwater longer than any other mammals except for whales and dolphins. Penguins lost their ability to fly in transforming their wings into strong fins.

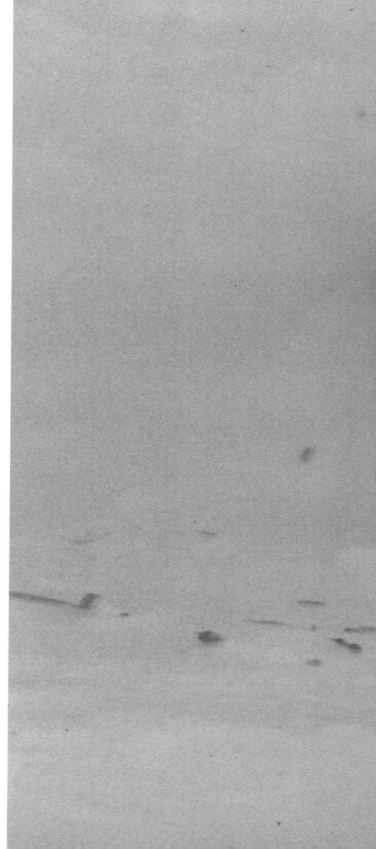

Only one species of penguin, the emperor penguin, manages to live and reproduce on the Antarctic continent in midwinter. It has to withstand almost unbearable conditions. This choice has enabled emperor penguins to protect their young from the risks of predation which are always present in the other colonies of penguins reproducing nearer the water. Fossils date the first penguins back at least 45 million years; some earlier species reached a height of five feet and weighed as much as 290 pounds.

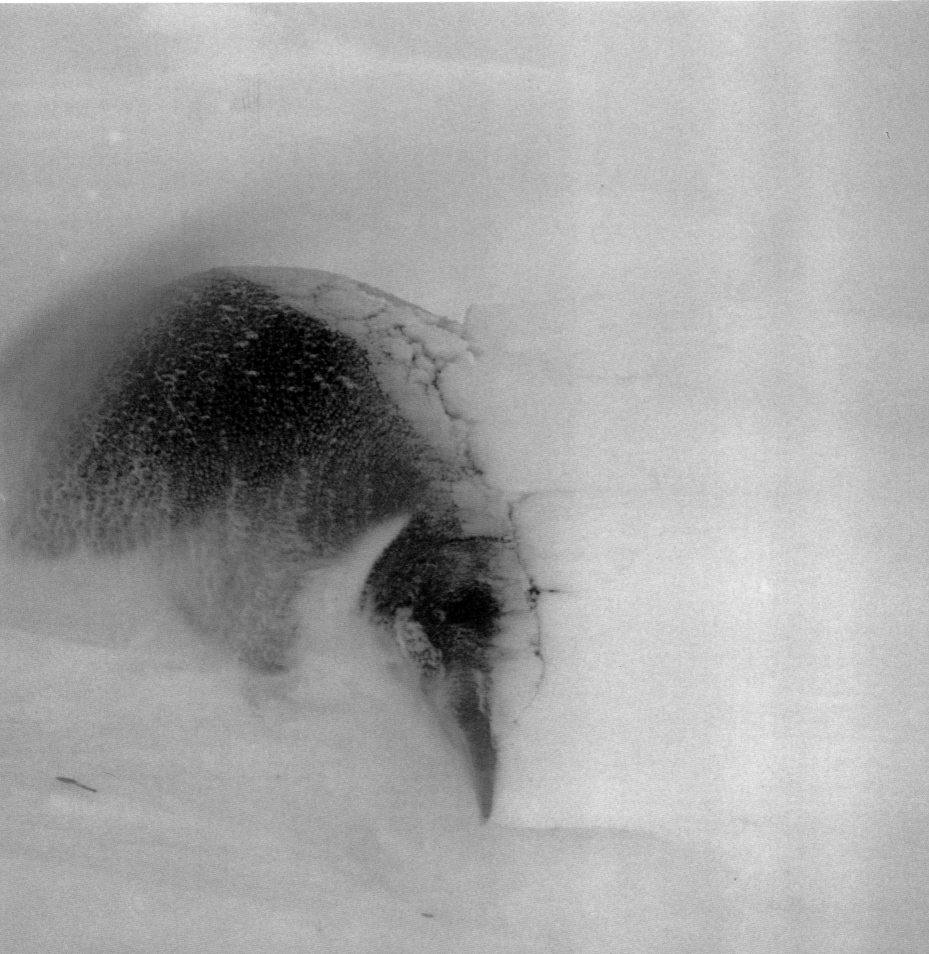

Polar bears are the only bears having a diet consisting exclusively of meat—their environment has nothing better to offer. They depend almost entirely on seals they capture as they roam about the Arctic. It is highly unlikely that polar bears ever moved much further south, even during the ice ages, as seals only live in ice-covered zones near the sea. Fossil remains of the first bears, similar to brown bears but smaller, have been found dating back to the Oligocene epoch. Since then, bears have spread throughout the world. Modern-day temperate zones were polar environments during the ice ages, and it was then that huge prehistoric cave bears fell prey to the first men coming from the Middle East, who eventually succeeded in extinguishing the species.

PAGES 114–115
Antarctica

CHAPTER FIVE

Forests

Perhaps trees have lost the great contest of evolutionary efficiency. Quicker than trees in the diffusion of their seeds, more resistant to environmental stresses, grasses provide a better response to the requirements of most environments. But for most of us, forests have an attraction which is different, perhaps more complete, than the prairies.

Every territory has an ecosystem that represents the arrival point of its history. Such an ecosystem contains the maximum number of animal or vegetable species, and the interrelationships and energy exchanges among the various populations are highly organized and infinitely complex. Ecologists call this stage climax. For the majority of environments, climax is represented by the formation of tall trees of different ages, a good covering of undergrowth, and a wealth of highly developed and closely interconnected animal species. In short, it is a forest.

The world's forests have very little in common apart from the preponderant presence of trees. A gloomy Nordic taiga does not have even one species similar to the trees of tropical forests, and a mediterranean wood shares no common environmental characteristics with a North American sequoia forest.

The arboreal form is not a recent conquest of the plant kingdom. There were already large expanses of forest in the Upper Devonian period. The Carboniferous period, of course, is named for the product of the plants that thrived during this era; in no other geological period are plant fossils so conspicuous. The deposits of coal we exploit today are the accumulations of materials that were created between 360 and 280 million years ago. If we bear in mind that it takes several cubic yards of living plants to produce a cubic yard of coal, we can imagine the enormous mass of vegetation present in the swamps of the Carboniferous period.

These swamps were dominated by gigantic tree ferns and club mosses, ancestors of the much more modest horsetails and ferns living today in the undergrowth. In the course of time, other more evolved forms gained the upper hand. These ranged from conifers like the primitive walchia which developed in the Permian period, to angiosperms, which originated in the Cretaceous.

In any case, primitive or evolved, conifers or angiosperms, the life in and around these trees was characterized, as in modern forests, by great complexity. This is perhaps the one thing all forests have in common—three-dimensionality; the addition of the dimension of height to the ecosystem enables a large number of other species to find their own niche.

It was from his observations of the life spreading up the trunk of a tree that Robert MacArthur, one of the greatest modern ecologists, was able to formulate the concept of the ecological niche. After many thousands of hours of research MacArthur was able to demonstrate that five different species of birds of the Dendroica family were capable of surviving on one tree. One feeds on the thin branches at the top of the tree, another on the ground beneath the tree, and yet another looks for its food under the pine needles. The ecological niche was thus defined as the role an organism plays, as its function within an ecosystem, or the "work" of a species.

Forests, along with some marine habitats, are where ecological niches are most numerous, most varied, and most minutely subdivided. Consequently, a greater number of species live in these habitats. To become suitable for providing hospitality to a large number of plants and animals in close relationship with one another, the ecosystem must be stable for as long a time as possible. As we have seen, glaciation or desertification, with the consequent changes in climatic and environmental conditions, do not aid stability or increase the complexity of interrelations among the species. For these historical reasons and also because of present-day limitations, some forests are no longer so complex and extraordinarily rich as one might expect. The number of species does tend to increase, however, when we move from the poles toward the equator.

The first true forests we encounter are the enormous stands of conifers in the North American and Eurasian taiga. As we saw in the chapter dedicated to the mountains, of which the taiga duplicates some conditions, extreme environments are dominated by conifers, which can best withstand long periods without photosynthesis or much water. In fact, the permanent frost greatly reduces

Gorillas, the largest living apes, only survive in three very small areas of equatorial Africa, and their number is in constant decline. Studies have established that the genetic differences between man and gorilla are minimal, perhaps no more than three percent—even less than the difference between man and chimpanzee. This means that the change in a very few genes was sufficient to transform a forest ape into an intelligent savanna biped.

the humidity of the air and makes the climate almost as arid as that of a desert. The leaves must therefore waste as little water as possible. Even at a lower level of productivity, no broadleaf tree could survive in this environment. In some areas the winters are so harsh that even conifer needles can be damaged. In this case, almost always, only larch trees are present, as these are one of the few gymnosperms losing their leaves in winter.

The great difference between summer and winter prevents a large number of animal species from surviving all year round in the taiga. In winter, only the extraordinary Siberian jay and the Lapland owl can survive here. This latter species exemplifies one of the strategies of the inhabitants of coniferous forests. When large numbers of prey are available, the brood can contain up to five young; when there is a shortage of small rodents in the forest, the owl might not reproduce at all. There are few insectivores, and they only stay a few weeks in the taiga, quickly moving further south as soon as the weather worsens. Mammals can reach large dimensions, and this helps them conserve heat. Bears, wolves, and wolverines are among the largest predators, but there are also the small and extremely quick martens, which prey on the abundant rodents, as well as some tiny Strigiformes (like the dwarf owl or the large-headed owl) which manage to capture rats and field mice as big as they themselves.

In milder latitudes, winters are shorter, and conifers are replaced by more adaptable angiosperms, or flowering plants and trees. Temperate forests grow thick with a great variety of species. An immense wooded canopy welcomed the first humans who penetrated into Europe, and later, the settlers who disembarked in North America. Despite the great diversity of latitudes in which it grows, the temperate forest has a similar structure in all parts of the globe. A curious aspect of the ecology of the temperate forest is that in the southern hemisphere the same latitudes are dominated by forests of evergreens. Perhaps as a consequence of long periods of drought, only the species in the northern hemisphere managed to evolve.

Temperate forests are not a single and monotonous formation. They include the so-called evergreen forests of which the Mediterranean maquis is the best known. Now nothing but a faint memory of what it was before the arrival of the Greeks and the Romans, the maquis is a fundamental crossroads for flora and fauna. Species coming from Africa and Asia, as well as plants adapted to dry and wet climates, encounter each other in this small corner of the Earth. The most evolved state, the climax, is a forest of evergreen trees that has adapted to a dry climate. The hottest period of the year brings a persistent drought, which leathery leaves enable these plants to withstand.

As one approaches the tropics, the forest loses the character of seasonality. Continuous rain and warmth transforms the forest into something unique and extraordinary which only in the last few years have we learned to appreciate as the richest environment on the surface of the Earth. Tropical forests completely cover all surfaces except the jungle rivers, and sometimes, as happens in some parts of Africa, they even spread several yards into the water. As in coral reefs, everything here is conducive to the full development of life. An abundance of light is equally distributed throughout the year, nourishment is rapidly recycled, and the temperature never descends below a comfortable range. On the equatorial belt, in South America, Africa, and Southeast Asia, there are real monuments to the creativity of nature, much superior in importance and complexity to any work of man. Seen from above, tropical forests look like an unbroken carpet of leaves. Beneath the canopy, there are at least another four or five layers of trees.

A short account cannot convey the complexity and interdependence existing in these forests. Rain forests have the greatest wealth of species on Earth. Although they occupy less than six percent of the Earth's surface, they contain between twenty and thirty percent of known animal and plant species. Our understanding of the dynamics of the tropical forests has greatly increased in recent years, but even today, many scientists still explore such important matters as the surface area, number of species, and the evolution of the forests themselves. They represent the Earth's ultimate example of biocenosis, or interactive self-sufficiency, and are perhaps the Earth's most evolved places.

Tropical forests are not extremely old, as was once thought— it seems that during the last ice

age, they were much less extensive than today. But they have permitted the development and evolution of an extremely impressive number of species, often unique and endemic. It is only here that we can see the most bizarre and colorful species, especially among amphibians and reptiles. Such unbridled fantasy arose out of an obscure past; fossils are very rare in these places where everything decomposes rapidly. Yet we have discovered in the fossil clays a meganeura, a gigantic dragonfly with a wing span of over two feet, and oversized Jurassic cockroaches, scorpions, spiders, and centipedes. Before the forests were inhabited by vertebrates, insects functioned as prey and predator. Even today the world of insects in the tropical forests has an importance perhaps superior to that of the animals—just consider the hordes of ants marching across the jungle floor or constructing elaborate nests high in the trees.

There were no extremely large animals in the forests of the past in the same way as there are none nowadays. We can, however, imagine that the luxuriant vegetation and the high production of flowers and fruit would have attracted a large number of animals to the edges of the forests. The enormous styracosaurus sought refuge from predators and nourishment in forest clearings. Just as male deer fight to establish a hierarchy, perhaps the clashing of the horns of these gigantic herbivores resounded in the clearings of ancient forests.

After the disappearance of the dinosaurs, reptiles became a minor component of the Earth's fauna. However, in the forests their numbers and variety still rival that of mammals and birds. On a small scale, basilisks have no reason to envy their ancient cousins. Their armor-plating and their crests, which are larger in the male, are similar to those of a sphinosaurus, one of the most active and ferocious predators of the Cretaceous period. Equally primeval-looking iguana scurry along the branches of tropical forests. Huge crocodiles infest the rivers. In the jungle, it is these large and small monsters that really remind us of primordial times.

Another numerous and widely dispersed group found in the forests is the serpents. The loss of their legs turned out to be very useful for speedy and silent movement across the forest floor. In an environment where sight is sometimes of little use,

snakes have an extraordinary heat perception mechanism capable of discerning even a difference of a few tenths of a degree, and they can thus follow their prey closely in absolute darkness.

As for birds, it was in a Jurassic forest that the well-known archaeopteryx took off for its first hesitant flight. We can get an idea of what this ancestor of all birds was like by observing the awkward flight of a hoatzin in the dense forests of Peru. The young of this strange bird have minuscule claws on their wings which are useful for climbing and fleeing from numerous predators. The hoatzin is not a primitive bird—in a sense, it has never lost some of the adaptations that were useful for the archaeopteryx.

To try to judge the wealth of life in the forests from the few fossils we have is like trying to count animals in a midwinter field to obtain an idea of the diversity of life. On the other hand, the upper layer alone of the modern-day forest may contain up to 30 million species, and that is only taking insects into account. To each of these species dozens of others can be connected in a maze of links and interrelationships difficult to decipher. Why is there such a wealth of life? As is the case with the sea, the important thing in an ecosystem is not so much the size as the number of ecological niches which exist. In their turn, these are roughly determined by the number of physical environments created by the vegetation.

In the forests there are at least five layers of vegetation; it is as if there were four ecosystems one on top of the other, each of which provides hospitality for a series of plant and animal communities completely different from one another. Beneath the canopy, rich in light, the others only receive a few rays of sun. At ground level, growth is slow but constant because environmental variations are almost nonexistent. The masters in this layer are mushrooms and animals living underground, extracting nourishment from the dead material which continuously falls from above. On each of the larger branches there are tens of other species making use of the gigantic trunk to reach the light. Each of these epiphytes, nonparasitic plants which live on a larger plant, in its turn can be host to tens of other smaller species. A small, colorful frog may spend its entire life on the leaves of a bromeliad in

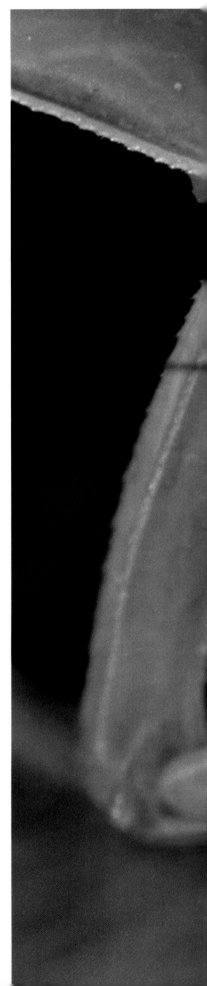

The world of the insects is represented in all the forests of the world by thousands, perhaps millions, of species. Among the most widespread groups are the butterflies, such as the Australian moth (upper right), and the mantids, such as the European praying mantis *shown devouring her mate (lower right). Because of their small and delicate bodies, it is not common to find remains of insects dating back to remote epochs. One of the principal sources of such finds is amber, a fossil tree resin which often trapped prehistoric insects.*

the Brazilian forest without ever seeing the ground and without having any necessity to do so.

One layer in particular, the canopy, about 100 feet from the ground, gives shelter to the majority of forest life: birds, insects, monkeys, predators, and flowering plants. The presence of so many different species creates a very large biomass. Although some estimates made in the 1970s have been corrected because they were too high, it would seem that the average biomass is about 176 tons for every 100 acres.

Tropical forests can be divided into at least four regions: American, African, Indo-Malayan, and Australasian. Each of them has different species which dominate different habitats and which distribute resources according to different dynamics. This means that the ways of making ends meet in the forest are much more varied than in any other environment. Most species in a tropical forest are more colorful and bizarre than the animals we are accustomed to seeing. This may be due to the wealth of available nutrients, which has permitted evolution to satisfy its creative whims. No longer forced to be constantly searching for food, many animals have managed to use some of their excess energy to become better equipped for the battle between the species as well as between the males of each species. Many extraordinary ornaments, beaks, calls, and colors are simply an extra strategy for attracting females. In the forests of New Guinea live some birds which have been described as nature's greatest artists: the male garden bird builds an extremely complex nest with no aim of providing shelter, while in the branches above, the bird of paradise shows off its extraordinary feathers to attract its own females.

Forests, the cradle of our species, the laboratories of evolution, are threatened by man like no other environment. Our not-too-distant cousins, the anthropoid apes (gorillas, chimpanzees, and orangutans) are still forest animals. In dense Asian and South American forests there still live some of the most beautiful and powerful predators on Earth: tigers, leopards, and jaguars. The world's tropical forests are perhaps nature's highest forms. Will the human species show itself to be worthy of the Earth by protecting these precious places?

Very large and very small animal forms have often developed on small islands far from the mainland. The most impressive example of this is the monitor lizard of Komodo, which lives in a few islands of the Sonda Archipelago. Its gigantic size (over nine feet long and weighing about 200 pounds) makes it a reliable model, if not of the true dinosaurs, at least of the most primitive ancestors of these animals which dominated the Earth during the Mesozoic. The Varanidae, to which the Komodo monitor lizard belongs, are sturdy animals which have some characteristics similar to the serpents. Millions of years ago, before the appearance of mammals, they were the largest predators on the islands, as they are today.

Scenes like this cannot help but recall the prehistoric era. The raised scales of the monitor lizard are almost identical to those of a carnosaurus, a theropod dinosaur which lived in the lower Cretaceous period in Argentina. Its even teeth hark back to the predatory reptiles of the Mesozoic. Its large claws are probably not very different from those of a coeleophysis, a fast-moving predator which reached a length of ten feet. The absence of large carnivorous mammals permitted the evolution and the survival of this peculiar species.

The colors of South American basilisks and vine snakes (below) *mimic the greens of the forest. Only many millions of years after plants conquered the land did they create complex ecological networks. One such tropical forest is on Mindanao Island in the Philip-* pines (below, right). *Shortly after the arrival of the green plants, the first arthropods, similar to herbivorous scorpions, clambered onto the land. It was perhaps to follow these that some fish transformed themselves into amphibians. Finally these latter freed them-* selves completely from their dependence on the water when they developed the terrestrial egg and became reptiles. It is not improbable that the group which eventually diversified to become mammals is as old as the reptiles.

PAGES 134–135
Bengal tiger, Ranthambor, India

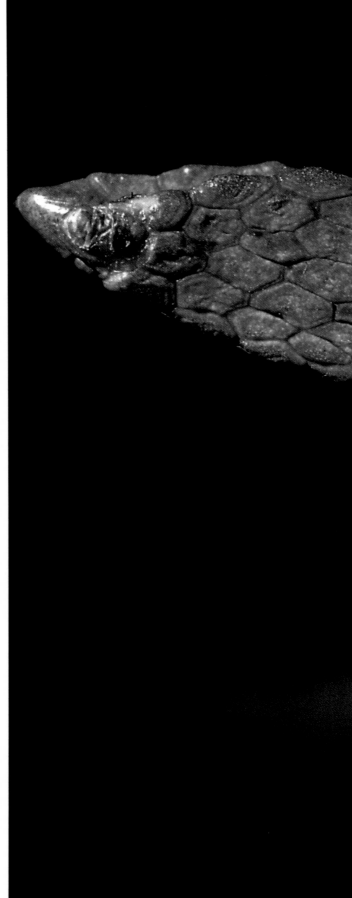

Chameleons are found all over Africa, but the strangest ones are found in Madagascar, where there are several species with horns and frontal protuberances. Islands which break away from the principal continents often remain isolated for millions of

years and are inhabited by peculiar animals and plants. Examples of this phenomenon are Madagascar and Australia. The absence of groups of large predators on these "lost worlds" enables others to evolve and substitute for them in their ecological function.

A unique ecosystem exists in the trees of tropical forests. Pictured at left is a rain forest canopy in Costa Rica. Many arboreal species, such as South America's sloth and New Guinea's bird of paradise (below), never come into con- tact with the ground. Watching a young sloth climbing upside down, one cannot help think- ing that until six million years ago bear-sized terrestrial sloths roamed over the South Ameri- can plains.

PAGES 140-141

Langurs on a banyan tree, Ranthambor, India

PAGES 142-143

Young orangutan, Indonesia

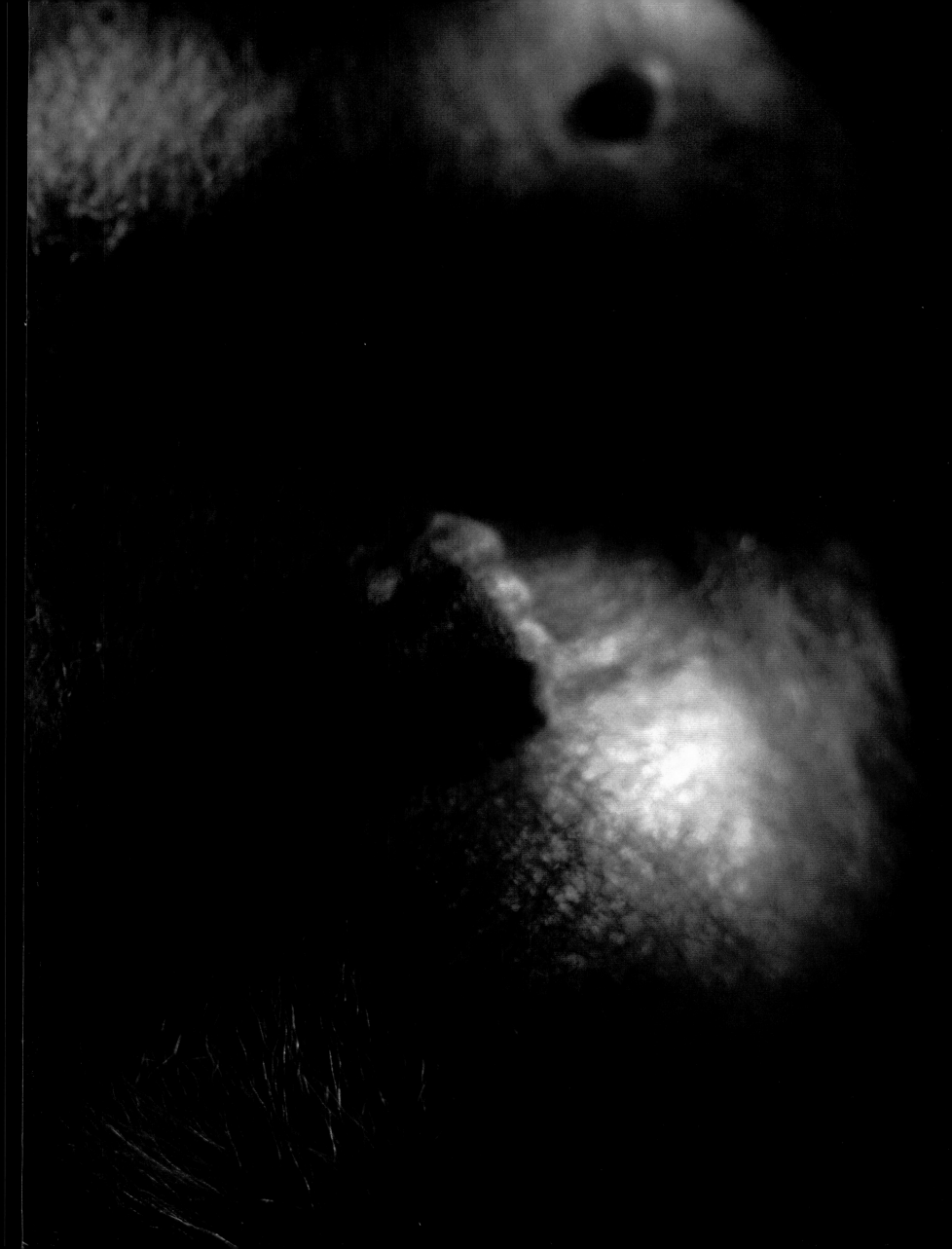

Photography Credits

Cover: Frans Lanting/Zefa
Back cover: Günter Ziesler

Kelvin Aitken: *pages 6-7, 32-33, 43 (all 3 photographs).*
D. Allan/Panda Photo: *page 109.*
Kurt Amsler/Planet Earth: *page 65 top.*
Alessandro Bardi/Panda Photo: *page 117.*
Jean and Des Bartlett/Bruce Coleman: *page 85.*
Andre Bartschi/Bruce Coleman: *pages 58-59.*
Erwin and Peggy Bauer: *pages 102-103, 103.*
Berol Cinematograph/Shot Photo: *pages 38 top, 64-65.*
Marcello Bertinetti: *pages 36-37, 62-63, 82-83, 99, 104-105, 106-107, 126-127, 128-129 (all 3 photographs), 130-131 (all 3 photographs), 138-139.*
Bildarchiv M. Harvey/Panda Photo: *page 78 top.*
M. Boulton/Panda Photo: *pages 88-89.*
J. R. Brackgirdck/Planet Earth: *pages 90-91.*
Jim Brandenburg/Planet Earth: *page 74.*
Bojan Brecelj and Arne Hodalic: *pages 24-25.*
Jane Burton/Bruce Coleman: *pages 54, 118.*
A. Calegari/Panda Photo: *page 136 bottom.*
John Cancalosi/Bruce Coleman: *pages 79, 80-81.*
Giuliano Cappelli/Panda Photo: *pages 4-5.*
Giuliano Colliva: *pages 84-85.*

Emanuele Coppola/Panda Photo: *pages 136-137.*
Nigel Dennis/Antony Bannister Photolibrary: *pages 96-97.*
Nicholas Devore III/Bruce Coleman: *page 78 bottom.*
GiPi Dore/Aster Italia: *page 42-43.*
J. Dragesco/Panda Photo: *page 14.*
Andrea Ferrari/Overseas: *pages 2-3.*
Jean-Paul Ferrero/Ardea London: *pages 142-143.*
Michael Fogden/Bruce Coleman: *page 132 top.*
Michael Fogden/Grazia Neri: *page 139 top.*
Foley/Zefa: *pages 60-61.*
Jeff Foott/Bruce Coleman: *page 144.*
Michael Freeman/Bruce Coleman: *pages 72-73.*
E. R. Gargiulo/Shot Photo: *page 19.*
M. P. Kahl/Bruce Coleman: *page 90.*
M. Lijima/Ardea London: *pages 27, 28-29, 68-69, 69 top.*
David Maitlend/Planet Earth: *page 123 top.*
Marco Mecklenburg: *page 44.*
McDougal Tiger Tops/Ardea London: *page 69 bottom.*
Rolando Menardi/Panda Photo: *page 21.*
WWF-F. Mercay/Panda Photo: *page 87.*
P. Morris/Ardea London: *pages 120-121.*
Mark Nissen: *pages 34-35, 46.*
M. Oggioni/Panda Photo: *page 112 top, 122-123.*
Osmond/Ardea London: *pages 12-13.*
Vincenzo Paolillo: *pages 30, 38 bottom, 38-39.*

D. Parer and E. Parer Cook/ Ardea London: *pages 70-71, 108, 114-115, 116.*
Daniele Pellegrini: *page 100.*
L. Piazza/Panda Photo: *page 51.*
R. Price/Grazia Neri: *page 111 top.*
Luciano Ramires: *page 98.*
Armstrong Roberts/Zefa: *pages 22-23.*
Graham Robertson/Ardea London: *pages 110, 110-111.*
Jeffrey L. Rotman: *pages 16-17, 48-49, 50.*
N. Rosing: *pages 26, 112-113, 113.*
Jonathan Scott/Planet Earth: *pages 66, 67, 76-77, 92-93.*
Peter Scoones/Planet Earth: *page 31.*
Massimo and Lucia Simion: *page 64 bottom.*
Marty Snyderman/Planet Earth: *page 45.*
Ron and Valerie Taylor/Ardea London: *page 18.*
Joe Van Warner/Bruce Coleman: *page 86.*
Adrian Warren/Ardea London: *pages 8, 9.*
James D. Watt: *page 47.*
Michael Melford Wheeler/Grazia Neri: *page 132 bottom.*
Richard Woldendorp: *pages 15, 55, 75.*
Rodney Wood/Planet Earth: *page 64 top.*
Konrad Wother/Bruce Coleman: *page 136 top.*
Günter Ziesler: *pages 10-11, 52-53, 56-57, 94-95 (all 3 photographs), 124-125, 132-133, 134-135, 139 bottom, 140-141.*
Claudio Ziraldo: *pages 40-41.*